A JOURNAL FROM OUR LEGATION
IN BELGIUM

Photograph by Boute, Brussels

His Majesty, Albert, King of the Belgians

A JOURNAL
FROM OUR LEGATION
IN BELGIUM

BY

HUGH GIBSON

SECRETARY OF THE AMERICAN LEGATION
IN BRUSSELS

Illustrated from photographs

DOUBLEDAY, PAGE & COMPANY
GARDEN CITY   NEW YORK
1917

TO MY MOTHER

LIST OF ILLUSTRATIONS

vii

LIST OF ILLUSTRATIONS

ix

INTRODUCTION

This volume is not a carefully prepared treatise on the war. It does not set out to prove anything. It is merely what its title indicates—a private journal jotted down hastily from day to day in odd moments, when more pressing duties would permit. Much material has been eliminated as of little interest. Other material of interest has been left out because it cannot be published at this time. It is believed, however, that what is printed here will suffice to give some idea of life in Belgium during the first few months of the war.

I have eliminated from the journal most of the matter about the early history of the Commission for Relief in Belgium. My day-to-day record did not do any sort of justice to the subject, and since it was not adequate, I have preferred to eliminate all but such casual reference to the relief work as is necessary to maintain the narrative. I am reconciled to this treatment of the subject by the knowledge that the story will be told comprehensively and well by Dr. Vernon Kellogg, who will soon publish an authoritative history of the Commission's work. As former Director of the Commission in Belgium, he has the detailed knowledge of its workings and the sympathetic understanding of its purpose, which peculiarly fit him for the task.

The work of the Commission is of a scope and significance that few of us realise. It is without doubt the greatest humanitarian enterprise in history, con-

xi

ducted under conditions of almost incredible difficulty. To those who had an understanding of the work, it had a compelling appeal, not only as an opportunity for service but also as the greatest conservation project of all time—the conservation of one of the finest races of our civilisation.

In its inception and execution, the work of the Commission is distinctively American. Its inception was in the mind of Herbert Hoover; in its execution he had the whole-hearted assistance of a little band of quiet American gentlemen who laboured in Belgium from the autumn of 1914 until we entered the war in April of this year. They came from all parts of our country and from all walks of life. They were simple work-a-day Americans, welded together by unwavering devotion to the common task and to Herbert Hoover, the "Chief." It was the splendid human side of the Commission that made it succeed in spite of all obstacles, and that part of the story will be hard to tell.

The gallant little band is now widely scattered. Some are carrying on their old work from Holland or England or America in order to ensure a steady flow of food to Belgium. Others are serving our Government in various capacities or fighting in the armies of our allies. Some of them we shall not see again and there will never be another reunion, as in the old days, when the "Chief" came over from London to Brussels with work to be done. But the bright light of kindly human service which brought them all together is still aflame and will always be an inspiration to those who served, however humbly, in the great work.

WASHINGTON, D. C., SEPT. 24, 1917.

A JOURNAL FROM OUR LEGATION
IN BELGIUM

A Journal From Our Legation In Belgium

*B*RUSSELS, *July 4, 1914.*—After years of hard work and revolutions and wars and rumours of war, the change to this quiet post has been most welcome and I have wallowed in the luxury of having time to play.

For the last year or two I have looked forward to just such a post as this, where nothing ever happens, where there is no earthly chance of being called out of bed in the middle of the night to see the human race brawling over its differences. When pounding along in the small hours of the night, nearly dead with fatigue, I have thought that I should like to have a long assignment to just such a post and become a diplomatic Lotos Eater. And at first it was great fun.

That phase lasted until I had had a thorough rest, and then the longing for something more active began to manifest itself.

I sat down and wrote to the Department of State that while I greatly appreciated having been sent to this much-coveted post I was ready whenever there might be need of my services to go where there was work to be done.

July 28, 1914.—Well, the roof has fallen in. War was declared this afternoon by Austria. The town is seething with excitement and everybody seems to realise how near they are to the big stage. Three classes of reserves have already been called to the colours to defend Belgian neutrality. A general mobilisation is prepared and may be declared at any time. The Bourse has been closed to prevent too much play on the situation, and let things steady themselves. In every other way the hatches have been battened down and preparations made for heavy weather.

To-night the streets are crowded and demonstrations for and against war are being held. The Socialists have Jaurés, their French leader, up from Paris, and have him haranguing an anti-war demonstration in the Grande Place, where a tremendous crowd has collected. Nobody on earth can see where it will all lead. England is trying hard to localise the conflict, and has valuable help. If she does not succeed * * *

An advance guard of tourists is arriving from France, Germany, and Switzerland, and a lot of them drop in for advice as to whether it is safe for them to go to various places in Europe. And most of them seem to feel that we really have authoritative information as to what the next few days are to bring forth, and resent the fact that we are too disobliging to tell them the inside news. A deluge of this sort would be easier for a full-sized Embassy to grapple with, but as Belgium is one of those places where nothing ever happens we have the smallest possible organisation, consisting on a peace basis of the Minister and myself, with one clerk. We shall have somehow to build up an emergency force to meet the situation.

July 30th.—No line on the future yet. Brussels is beginning to look warlike. Troops are beginning to appear. The railway stations have been occupied, and the Boy Scouts are swarming over the town as busy as bird dogs. A week ago there was hardly a tourist in Brussels. Now the Legation hall is filled with them, and they all demand precise information as to what is going to happen next and where they can go with a guarantee from the Legation that they will not get into trouble.

July 31st.—No, my recent remarks about nothing ever happening in Brussels were not intended as sarcasm. I thought Belgium was the one place where I could be sure of a quiet time, and here we are right in the centre of it. Even if nothing more happens we have had enough excitement to last me for some time. The doings of the past few days have brought out some idea of what a general European war would mean—and it is altogether too dreadful to think of.

Saturday, Aug. 1st.—Last night when I went home, at about midnight, I found the police going about with the orders for mobilisation, ringing the door bells and summoning the men to the colours. There was no time to tarry, but each man tumbled out of bed into his clothes and hurried away to his regiment. Two of my neighbours were routed out a little after midnight, and got away within the hour. There was a good deal of weeping and handshaking and farewelling, and it was not the sort of thing to promote restful sleep.

This morning I got down to the chancery at a quarter past eight, and found that Omer, our good messenger, had been summoned to the colours. He had gone, of course, and had left a note for me to announce the fact. He had been ill, and could perfectly well have been exempted. The other day, when we had discussed the matter, I had told him that there would be no difficulty in getting him off. He showed no enthusiasm, however, and merely remarked, without heroics, that it was up to him.

Colonel Falls, 7th Regiment, of the National Guard of New York, came in, having been sent back from the frontier. He had the pleasure of standing all the way as the trains were packed.

Millard Shaler, the American mining engineer, who had just come back from the Congo, came in with his amusing Belgian friend who had been telling us for weeks about the wonderful new car in which he was investing. This time he came around to let me have a look at it, he having been advised that the car was requisitioned and due to be taken over to-day.

We have done a land-office business in passports, and shall probably continue to turn them out by the dozen.

Sunday, August 2d.—Another hectic day with promise of more to come.

This morning I came down a little earlier than usual and found the Minister already hard at it. He had been routed out of bed and had not had time to bathe or shave. There was nothing to show that it was a Sunday—nearly twice as many callers as yesterday, and they were more exacting.

Mrs. A— B— C— came in airily and announced that she had started from Paris yesterday on a motor tour through France and Belgium. Having got this far, some rude person had told her that her motor might be seized by the Government for military purposes and that an order had been promulgated forbidding any one to take cars out of the country. She came around confidently to have us assure her that this was a wicked lie—and needless to say was deeply disappointed in us when we failed to back her up. We had refrained from asking the Government to release our own servants from their military obligations and have refused to interfere for anybody else, but that was not enough for her. She left, a highly indignant lady.

The story is around town this afternoon that the Germans have already crossed the frontier without the formality of a declaration of war—but that remains to be seen. Brussels was put under martial law last night, and is now patrolled by grenadiers and lancers.

The money situation is bad. All small change has disappeared in the general panic, and none of it has dared show its head during the past few days. The next thing done by panicky people was to pass round word that the Government bank notes were no good and would not be honoured. Lots of shops are refusing to accept bank notes, and few places can make any change. The police are lined up outside the banks keeping people in line. People in general are frantic with fear, and are trampling each other in the rush to get money out of the banks before the crash that probably will not come. Travelers who came here with pockets bulging with express checks and bank notes are unable to get a cent of real money, and nobody shows any enthusiasm over American paper. I have a few bank notes left, and this evening when I went into a restaurant I have patronised ever since my arrival the head waiter refused to change a note for me, and I finally had to leave it and take credit against future meals to be eaten there. We may have our troubles when our small store is gone, but probably the situation will improve and I refuse to worry. And some of our compatriots don't understand why the Legation does not have a cellar full of hard money to finance them through their stay in Europe.

Communications, with such parts of the world as we still speak to, are getting very difficult on account of mobilisation, the military having right of way. This morning's Paris papers have not come in this evening, and there are no promises as to when we shall see them. The news in the local papers is scarce and doubtful, and I hope for a word from Paris.

Word has just come in that the Government has seized the supplies of bread, rice, and beans, and will fix prices for the present. That is a sensible and steadying thing, and should have a good effect.

Nobody seems to remember that a few days ago Serbia was playing a star rôle in this affair. She seems to have faded away behind the scenes. A few days ago, Mexico loomed large in the papers and now we have forgotten that she ever existed. Albania supplied a lot of table talk, and now we think about as much about her and her troubles as we do about Thibet.

This afternoon I went around to the Rue Ducale to take a look at the French Legation. The tricolor was flying in the fresh breeze, and there was a big crowd outside cheering itself hoarse. It was made up of men who were called to the colors and were waiting to enroll themselves and get instructions as to where they should report for duty. The air was electric, and every now and then the military band struck up the Marseillaise and the crowd instantly became happily delirious. Some of them had been standing in the sun for hours waiting to get in and get their orders, but they were just as keenly responsive to the music and the mood of the crowd as anybody. All the crowd in the Legation had been working day and night for days, and was dead with fatigue; but, some way, they kept going, and managed to be civil and friendly when I had business with them. How they do it I don't know. A Frenchman's politeness must be more deeply ingrained than even I had supposed.

On the way back from the Legation this evening, I saw von Below, the German Minister, driving home from the Foreign Office to his Legation. He passed close to me, and I saw that the perspiration was standing out on his forehead. He held his hat in his hand and puffed at a cigarette like a mechanical toy, blowing out jerky clouds of smoke. He looked neither to left nor right, and failed to give me his usual ceremonious bow. He is evidently not at ease about the situation, although he continues to figure in the newspapers as stating that all is well, that Germany has no intention of setting foot on Belgian soil, and that all Belgium has to do is to keep calm. In an interview given to

Le Soir he sums up his reassuring remarks by saying: "Your neighbour's house may burn but yours will be safe."

August 3, 1914.—No mail in to-day. All communications seem to be stopped for the time being at least. Mobilisation here and in France requires all the efforts of all hands, and little workaday things like mail and newspapers go by the board.

According to the news which was given me when I got out of bed this morning, the German Minister last night presented to the Belgian Government an ultimatum demanding the right to send German troops across Belgium to attack France. He was evidently returning from this pleasant duty when I saw him last night, for the ultimatum seems to have been presented at seven o'clock. The King presided over a Cabinet Council which sat all night; and when the twelve hours given by the ultimatum had expired, at seven this morning, a flat refusal was sent to the German Legation. Arrangements were got under way, as the Council sat, to defend the frontiers of the country against aggression. During the night the garrison left and the Garde Civique came on duty to police the town.

The influx of callers was greater to-day than at any time so far, and we were fairly swamped. Miss Larner came in and worked like a Trojan, taking passport applications and reassuring the women who wanted to be told that the Germans would not kill them even when they got to Brussels. She is a godsend to us.

Monsieur de Leval, the Belgian lawyer who for ten years has been the legal counselor of the Legation, came in and brought some good clerks with him. He also hung up his hat and went to work, making all sorts of calls at the Foreign Office, seeing callers, and going about to the different Legations. Granville Fortescue came in from Ostend, and I should have put him to work but that he had plans of his own and has decided to blossom forth as a war correspondent. He is all for getting to the "front" if any.

Just to see what would happen, I went to the telephone after lunch and asked to be put through to the Embassy at London. To my surprise, I got the connection in a few minutes and had a talk with Bell, the Second Secretary. The Cabinet had been sitting since eleven this morning, but had announced no decision. I telephoned him again this evening and got the same reply. Bell said that they had several hundred people in the chancery and were preparing for a heavy blow.

As nearly as we can make out the Germans have sent patrols into Belgian territory, but there have been no actual operations so far. All day long we have been getting stories to the effect that there has been a battle at Visé and that fifteen hundred Belgians had been killed; later it was stated that they had driven the Germans back with heavy losses. The net result is that at the end of the day we know little more than we did this morning.

Parliament is summoned to meet in special session to-morrow morning to hear what the King has to say about the German ultimatum. It will be an interesting sight. Parliament has long been rent with most bitter factional quarrels, but I hear that all these are to be forgotten and that all parties, Socialists included, are to rally round the throne in a great demonstration of loyalty.

All the regular troops have been withdrawn from this part of the country and dispatched to the front, leaving the protection of the capital to the Garde Civique, who are patrolling the streets, to examine the papers of everybody who moves about. This is a sort of local guard made up of people who have not been called for active military service, but who have volunteered for local defense. They are from every class—lawyers and butchers and bakers and dentists and university professors. They have, of course, had little training for this sort of work, and have had only elementary orders to guide them. These they carry out to the letter. There are detachments of them at all sorts of strategic points in the city where they hold up passing vehicles to see who is inside. I have been stopped by them goodness knows how many times this day. They hold up the car, look inside, apologise, and explain good-naturedly that they are obliged to bother me, asking who I am, and after I have satisfied them with papers that any well-equipped spy would be ashamed of, they let me go on with more apologies. They rejoice in a traditional uniform topped off by a derby hat with kangaroo feathers on it. This is anything but martial in appearance and seems to affect their funny bone as it does mine.

August 5th.—Yesterday morning we got about early and made for the Chamber of Deputies to hear the King's speech. The Minister and I walked over together and met a few straggling colleagues headed in the same direction. Most of them had got there ahead of us, and the galleries were all jammed. The Rue Royale, from the Palace around the park to the Parliament building, was packed with people, held in check by the Garde Civique. There was a buzz as of a thousand bees and every face was ablaze—the look of a people who have been trampled on for hundreds of years and have not learned to submit. The Garde Civique had two bands in front of the Senate, and they tried to play the Brabançonne in unison. Neither of them could play the air in tune, and they were about a bar apart all the time. They played it through and then began to play it over again without a pause between. They blew and pounded steadily for nearly half an hour, and the more they played, the more enthusiastic the crowds became.

When I saw how crowded the galleries were I thought I would not push, so resigned myself to missing the speech and went out onto a balcony with Webber, of the British Legation, to see the arrival of the King and Queen. We had the balcony to ourselves, as everybody else was inside fighting for a place in the galleries to hear the speech.

When the King and Queen finally left the Palace we knew it from a roar of cheering that came surging across the Park. The little procession came along at a smart trot, and although it was hidden from us by the trees we could follow its progress by the steadily advancing roaring of the mob. When they turned from the Rue Royale into the Rue de la Loi, the crowd in front of the Parliament buildings took up the cheering in a way to make the windows rattle.

First came the staff of the King and members of his household. Then the Queen, accompanied by the royal children, in an open daumont. The cheering for the Queen was full-throated and with no sign of doubt, because of her Bavarian birth and upbringing— she is looked on as a Belgian Queen and nothing else.

After the Queen came a carriage or two with members of the royal family and the Court. Finally the King on horseback. He was in the field uniform of a lieutenant-general, with no decorations and none of the ceremonial trappings usual on such occasions as a speech from the Throne. He was followed by a few members of his staff who also looked as though they were meant more for business than for dress parade.

As the King drew rein and dismounted, the cheering burst forth with twice its former volume; and, in a frantic demonstration of loyalty, hats and sticks were thrown into the air. Two bands played on manfully, but we could hear only an occasional discord.

Just as the King started into the building an usher came out, touched me on the arm and said something, beckoning me to come inside. One of the galleries had been locked by mistake but had now been opened, and Webber and I were rewarded for our modesty by being given the whole thing to ourselves. In a few minutes the Bolivian Chargé came in and joined us. Our places were not ten feet from the Throne, and we could not have been better placed.

The Queen came in quietly from one side and took a throne to the left of the tribune, after acknowledging a roaring welcome from the members of the two Houses. When the cheering had subsided, the King walked in alone from the right, bowed gravely to the assembly and walked quickly to the dais above and behind the tribune.   With a business-like gesture he tossed his cap on to the ledge before him and threw his white cotton gloves into it—then drew out his speech and read it.   At first his voice was not very steady but he soon controlled it and read the speech to the end in a voice that was vibrating with emotion but without any oratory or heroics.   He went straight to the vital need for union between all factions and all parties, between the French, Flemish, and Walloon races, between Catholics, Liberals, and Socialists in a determined resistance to the attack upon Belgian independence.   The House could contain itself for only a few minutes at a time, and as every point was driven home they burst into frantic cheering.   When the King, addressing himself directly to the members of Parliament, said, "Are you determined at any cost to maintain the sacred heritage of our ancestors?" the whole Chamber burst into a roar, and from the Socialists' side came cries of: "At any cost, by death if need be."

It was simple and to the point—a manly speech. And as he delivered it he was a kingly figure, facing for the sake of honour what he knew to be the gravest danger that could ever come to his country and his people.   When he had finished he bowed to the Queen, then to the Parliament, and then walked quickly out of the room, while the assembly roared again.   The Senators and Deputies swarmed about the King on his way out, cheering and trying to shake him by the hand—and none were more at pains to voice their devotion than the Socialists.

After he had gone the Queen rose, bowed shyly to the assembly, and withdrew with the royal children. She was given a rousing ovation as everybody realised the difficulty of her position and was doubly anxious to show her all their confidence and affection. The whole occasion was moving, but when the little Queen acknowledged the ovation so shyly and so sadly and withdrew, the tears were pretty near the surface—my surface at any rate.

For several minutes after the Queen withdrew the cheering continued. Suddenly a tense silence fell upon the room. M. de Broqueville, the Prime Minister, had mounted the tribune and stood waiting for attention. He was clearly under great stress of emotion, and as the House settled itself to hear him he brushed away the tears that had started to his eyes. He began in a very direct way by saying that he would limit himself to reading a few documents and hoped that, after hearing them, the House would consider the Government worthy of the confidence that had been reposed in it and that immediate action would be taken upon matters of urgent importance. He first read the German ultimatum,* which was received quietly but

---

* The following is the text of the German ultimatum:

BRUSSELS, August 2, 1914.

VERY CONFIDENTIAL.

Reliable information has been received by the German Government to the effect that French forces intend to march on the line of the Meuse by Gîvet and Namur. This information leaves no doubt as to the intention of France to march through Belgian territory against Germany.

The German Government cannot but fear that Belgium, in

Brüssel, den 2 August 1914

Très confidentiel.

Der Kaiserlichen Regierung liegen zuverlässige Nachrichten vor ueber den beabsichtigten Aufmarsch französischer Streitkräfte an der Maas-Strecke Givet-Namur. Sie lassen keinen Zweifel ueber die Absicht Frankreichs, durch belgisches Gebiet gegen Deutschland vorzugehen

Die Kaiserliche Regierung kann sich der Besorgniss nicht erwehren, daß Belgien, trotz besten Willens, nicht im Stande sein wird, ohne Hülfe einen französischen Vormarsch mit so großer Aussicht auf Erfolg abzuwehren, daß darin eine ausreichende Sicherheit gegen die Bedrohung Deutschlands gefunden werden kann Es ist ein Gebot der Selbsterhaltung für Deutschland, dem feindlichen Angriff zuvorzukommen. Mit dem größten Bedauern würde es daher die deutsche Regierung erfüllen, wenn Belgien einen Akt der Feindseligkeit

Fac-simile of the first page of the German ultimatum to Belgium

Pass issued by the Belgian military authorities to enable
Mr. Gibson to enter the German Legation at Brussels

Maître Gaston de Leval, legal adviser to the American Legation
in Brussels

spite of the utmost goodwill, will be unable, without assistance, to repel so considerable a French invasion with sufficient prospect of success to afford an adequate guarantee against danger to Germany. It is essential for the self-defence of Germany that she should anticipate any such hostile attack. The German Government would, however, feel the deepest regret if Belgium regarded as an act of hostility against herself the fact that the measures of Germany's opponents force Germany, for her own protection, to enter Belgian territory.

In order to exclude any possibility of misunderstanding, the German Government make the following declaration:

1. Germany has in view no act of hostility against Belgium. In the event of Belgium being prepared in the coming war to maintain an attitude of friendly neutrality towards Germany, the German Government bind themselves, at the conclusion of peace, to guarantee the possessions and independence of the Belgian Kingdom in full.

2. Germany undertakes, under the above-mentioned condition, to evacuate Belgian territory on the conclusion of peace.

3. If Belgium adopts a friendly attitude, Germany is prepared, in co-operation with the Belgian authorities, to purchase all necessaries for her troops against a cash payment, and to pay an indemnity for any damage that may have been caused by German troops.

4. Should Belgium oppose the German troops, and in particular should she throw difficulties in the way of their march by a resistance of the fortresses on the Meuse, or by destroying railways, roads, tunnels or other similar works, Germany will, to her regret, be compelled to consider Belgium as an enemy.

In this event, Germany can undertake no obligations towards Belgium, but the eventual adjustment of the relations between the two States must be left to the decision of arms.

The German Government, however, entertain the distinct hope that this eventuality will not occur, and that the Belgian Government will know how to take the necessary measures to prevent the occurrence of incidents such as those mentioned. In this case the friendly ties which bind the two neighbouring States will grow stronger and more enduring.

with indignation and anger which was with difficulty suppressed. Without commenting upon the German note, he then read the reply which had been handed to the German Minister.* This was followed by a final

---

*The Belgian Government replied as follows to the German ultimatum:

The German Government stated in their note of the 2nd August, 1914, that according to reliable information French forces intended to march on the Meuse via Gîvet and Namur, and that Belgium, in spite of the best intentions, would not be in a position to repulse, without assistance, an advance of French troops.

The German Government, therefore, considered themselves compelled to anticipate this attack and to violate Belgian territory. In these circumstances, Germany proposed that the Belgian Government adopt a friendly attitude towards her, and undertook, on the conclusion of peace, to guarantee the integrity of the Kingdom and its possessions to their full extent. The note added that if Belgium put difficulties in the way of the advance of German troops, Germany would be compelled to consider her as an enemy, and to leave the ultimate adjustment of the relations between the two States to the decision of arms.

This note has made a deep and painful impression upon the Belgian Government.

The intentions attributed to France by Germany are in contradiction to the formal declarations made to us on August 1st in the name of the French Government.

Moreover, if contrary to our expectation, Belgian neutrality should be violated by France, Belgium intends to fulfil her international obligations and the Belgian army would offer the most vigorous resistance to the invader.

The treaties of 1839, confirmed by the treaties of 1870, vouch for the independence and neutrality of Belgium under the guarantee of the Powers, and notably of His Majesty the King of Prussia.

note delivered by the German Minister this morning stating "that in view of the refusal of the King to accede to the well-intentioned proposals of the Emperor, the Imperial Government, greatly to its regret, was obliged to carry out by force of arms the measures indispensable to its security." After reading these documents he made a short and ringing speech, full of fire, which was repeatedly interrupted by cheers. When he came down from the tribune he was surrounded by cheering Senators and Deputies struggling to shake his hand and express their approval of his speech. Even the Socialists who had fought him for years rose to the occasion and vied with their colleagues in their demonstrations of enthusiasm. Broqueville rose again and said: "In the present crisis we have received from the opposition a whole-hearted support; they have

---

Belgium has always been faithful to her international obligations; she has carried out her duties in a spirit of loyal impartiality and she has left nothing undone to maintain and enforce respect for her neutrality.

The attack upon her independence with which the German Government threaten her constitutes a flagrant violation of international law. No strategic interest justifies such a violation of law.

The Belgian Government, if they were able to accept the proposals submitted to them, would sacrifice the honour of the nation and betray their duty towards Europe.

Conscious of the part which Belgium has played for more than eighty years in the civilisation of the world, they refuse to believe that the independence of Belgium can only be preserved at the price of the violation of her neutrality.

If this hope is disappointed the Belgian Government are firmly resolved to repel, by all the means in their power, every attack upon their rights.

Brussels, August 3, 1914 (7 A.M.).

rallied to our side in the most impressive way in preparing the reply to Germany. In order to emphasise this union of all factions, His Majesty the King has just signed a decree appointing Monsieur Vandervelde as a Minister of State." This announcement was greeted by roars of applause from all parts of the House, and Vandervelde was immediately surrounded by Ministers and Deputies anxious to congratulate him. His reply to the Prime Minister's speech was merely a shout above the roar of applause: "I accept."

As we came out, some of the colleagues were gathered about debating whether they should go over to the Palace and ask to take leave of the King. They were saved that labour, however, for the King had stepped into a motor at the door and was already speeding to the General Headquarters which has been set up nobody knows where. That looks like business.

When I got back to the Legation I found von Stumm, Counselor of the German Legation, with the news that his chief had received his passports and must leave at once. He had come to ask that the American Minister take over the care of the German Legation and the protection of the German subjects who had not yet left the country. I said that we could not undertake anything of that sort without authority from Washington, and got the Minister to telegraph for it when he came in from some hurried visits he had made in search of news.

While we were snatching some lunch, von Stumm came back with the German Minister, von Below, and said that some provisional arrangement must be made at once as the staff of the Legation would have to leave

for the Dutch frontier in the course of the afternoon—
long before we could hope for an answer from Washington. We did not like the idea of doing that sort of
thing without the knowledge of Washington, but
finally agreed to accept the charge provisionally on
grounds of humanity, until such time as we should
receive specific instructions as to who would be
definitely entrusted with the protection of German
interests. In case of need, we shall be asked to take
over certain other Legations and shall have our hands
more than full.

At five o'clock we went over to the German Legation, which we found surrounded by a heavy detachment of Garde Civique as a measure of protection
against violence. We drew up, signed, and sealed
a protocol accepting what is known as *la garde des
clefs et des sceaux*, until such time as definite arrangements might be made. The Minister and von Stumm
were nearly unstrung. They had been under a great
strain for some days and were making no effort to
get their belongings together to take them away.
They sat on the edge of their chairs, mopped their
brows and smoked cigarettes as fast as they could
light one from another. I was given a lot of final
instructions about things to be done—and all with the
statement that they should be done at once, as the
German army would doubtless be in Brussels in three
days. While we were talking, the chancellor of the
Legation, Hofrat Grabowsky, a typical white-haired
German functionary, was pottering about with sealing
wax and strips of paper, sealing the archives and
answering questions in a deliberate and perfectly calm

way. It was for all the world like a scene in a play. The shaded room, the two nervous diplomats registering anxiety and strain, the old functionary who was to stay behind to guard the archives and refused to be moved from his calm by the approaching cataclysm. It seemed altogether unreal, and I had to keep bringing myself back to a realisation of the fact that it was only too true and too serious.

They were very ominous about what an invasion means to this country, and kept referring to the army as a steam roller that will leave nothing standing in its path. Stumm kept repeating: "Oh, the poor fools! Why don't they get out of the way of the steam roller. We don't want to hurt them, but if they stand in our way they will be ground into the dirt. Oh, the poor fools!"

The Government had a special train ready for the German diplomatic and consular officers who were to leave, and they got away about seven. Now, thank goodness, they are safely in Holland and speeding back to their own country.

Before leaving, Below gave out word that we would look after German interests, and consequently we have been deluged with frightened people ever since.

All the Germans who have remained here seem to be paralysed with fright, and have for the most part taken refuge in convents, schools, etc. There are several hundreds of them in the German Consulate-General which has been provisioned as for a siege. Popular feeling is, of course, running high against them, and there may be incidents, but so far nothing has happened to justify the panic.

This morning a Belgian priest, the Abbé Upmans, came in to say that he had several hundred Germans under his care and wanted some provision made for getting them away before the situation got any worse.

After talking the matter over with the Minister and getting his instructions, I took the Abbé in tow, and with Monsieur de Leval went to the Foreign Office to see about getting a special train to take these people across the border into Holland and thence to Germany. At first, the suggestion was received with some resentment and I was told flatly that there was no good reason for Belgium to hand over special trains to benefit Germans when every car was needed for military operations. I pleaded that consideration must be shown these helpless people and that this course was just as much in the interest of Belgium as of anybody else, as it would remove the danger of violence with possible reprisals and would relieve the overworked police force of onerous duties. After some argument, Baron Donny went with me to the Sûreté Publique where we went over the matter again with the Chief. He got the point at once, and joined forces with us in a request to the Minister of Railways for a special train. We soon arranged matters as far as the Belgian frontier. I then telephoned through to The Hague, got Marshal Langhorne and asked him to request the Dutch Government to send another train to the frontier to pick our people up and send them through to Germany. He went off with a right good will to arrange that, and I hope to have an answer in the morning.

We plan to start the train on Friday morning at

four o'clock, so as to get our people through the streets when there are few people about. We are making it known that all Germans who wish to leave should put in an appearance by that time, and it looks as though we should have from seven hundred to a thousand to provide for. It will be a great relief to get them off, and I hold my breath until the train is safely gone.

The Belgian Government is making no distinction between Germans, and is letting those liable for military service get away with the others.

Wild stories have begun to circulate about what is bound to happen to Americans and other foreigners when hostilities get nearer to Brussels, and we have had to spend much time that could have been devoted to better things in calming a lot of excitable people of both sexes. I finally dug out the plan of organisation of the foreigners for the Siege of Peking and suggested to the Minister that, in order to give these people something to do and let them feel that something was being done, we should get them together and appoint them all on committees to look after different things. This was done to-day. Committees were appointed to look for a house where Americans could be assembled in case of hostilities in the immediate vicinity of Brussels; to look after the food supply; to attend to catering; to round up Americans and see that they get to the place of refuge when the time comes; to look after destitute Americans, etc. Now they are all happy and working like beavers, although there is little chance that their work will serve any useful purpose aside from keeping them occupied. We got

Mrs. Shaler to open up the Students' Club, which had been closed for the summer, so that the colony can have a place to meet and work for the Red Cross and keep its collective mind off the gossip that is flying about.

Last night our cipher telegrams to Washington were sent back from the telegraph office with word that under the latest instructions from the Government they could not be forwarded. The Minister and I hurried over to the Foreign Office, where we found several of the colleagues on the same errand. It was all a mistake, due to the fact that the General Staff had issued a sweeping order to stop all cipher messages without stopping to consider our special case. It was fixed after some debate, and the Minister and I came back to the shop and got off our last telegrams, which were finished at three this morning.

I was back at my desk by a little after eight and have not finished this day's work, although it is after midnight. I have averaged from three to five hours sleep since the trouble began and, strange to say, I thrive on it.

I have called several times to-day at the French and British Legations to get the latest news. They keep as well posted as is possible in the prevailing confusion, and are most generous and kind in giving us everything they properly can.

There seems to have been a serious engagement to-day at Liège, which the Germans are determined to reduce before proceeding toward France. The report is that the attack was well resisted and the Germans driven back with heavy loss. A number of prisoners have been taken and were being brought into Brussels this evening along with the wounded. In the course of the fighting there was a sort of charge of the Light Brigade; one squadron of Belgian Lancers was obliged to attack six times its number of Germans and was cut to pieces, only one officer escaping. The morale of the Belgians is splendid.

This afternoon as the Minister and I were going to call on the British Minister, we passed the King and his staff headed out the Rue de la Loi for the front. They looked like business.

Several times to-day I have talked over the telephone with the Embassy in London. They seem to be as strong on rumours as we are here. One rumour I was able to pass on to Bell was to the effect that the British flagship had been sunk by German mines with another big warship. Another to the effect that five German ships have been destroyed by the French fleet off the coast of Algeria, etc., etc.

The Red Cross is hard at work getting ready to handle the wounded, and everybody is doing something. Nearly everybody with a big house has fitted it in whole or in part as a hospital. Others are rolling bandages and preparing all sorts of supplies.

The military attachés are all going about in uniform now. Each Legation has a flag on its motor and the letters C. D.—which are supposed to stand for Corps Diplomatique, although nobody knows it. I have seized Mrs. Boyd's big car for my own use. D. L. Blount has put his car at the disposal of the Minister and is to drive it himself.

There is talk already of moving the Court and the Government to Antwerp, to take refuge behind the fortifications. When the Germans advance beyond Liège, the Government will, of course, have to go, and the diplomatic corps may follow. It would be a nuisance for us, and I hope we may be able to avoid it.

Germans are having an unhappy time, and I shall be happier when they are across the border. Nothing much seems to have happened to them beyond having a few shops wrecked in Antwerp and one or two people beaten up here. One case that came to my knowledge was an outraged man who had been roughly handled and could not understand why. All he had done was to stand in front of a café where the little tables are on the sidewalk and remark: "Talk all the French you can. You'll soon have to talk German." Of course there are a lot of Belgians, Swiss and Dutch who rejoice in good German names and they are not having a pleasant time. One restaurant called Chez Fritz, I saw when coming along the Boulevard this evening, had hung out a blackboard with the proud device: *"Fritz est Luxembourgeois, mais sa Maison est Belge."* He was taking no chances on having the place smashed.

August 6th.—This morning when I came into the Legation I found the Minister of Justice in top hat and frock coat waiting to see somebody. He had received a report that a wireless station had been established on top of the German Legation and was being run by the people who were left in the building. He came to ask the Minister's consent to send a judge to look, see and draw up a *procès verbal.* In our own artless little American way we suggested that it might

be simpler to go straight over and find out how much
there was to the report. The Minister of Justice had
a couple of telegraph linemen with him, and as soon
as Mr. Whitlock could get his hat, we walked around
the corner to the German Legation, rang the bell, told
the startled occupants that we wanted to go up to
the garret and—up we went.

When we got there we found that the only way onto
the roof was by a long perpendicular ladder leading
to a trap door. We all scrambled up this—all but the
Minister of Justice, who remained behind in the garret
with his top hat.

We looked the place over very carefully, and the
workmen—evidently in order to feel that they were
doing something—cut a few wires which probably
resulted in great inconvenience to perfectly harmless
people farther along the street. But there was no
evidence of a wireless outfit. One of the men started
to explain to me how that proved nothing at all; that
an apparatus was now made that could be concealed
in a hat and brought out at night to be worked. He
stopped in the middle of a word, for suddenly we heard
the rasping intermittent hiss of a wireless very near
at hand. Everybody stiffened up like a lot of pointers,
and in a minute had located the plant. It was nothing
but a rusty girouette on top of a chimney being turned
by the wind and scratching spitefully at every turn.
The discovery eased the strain and everybody laughed.

Then there was another sound, and we all turned
around to see a trap door raised and the serene, bemonocled face of my friend Cavalcanti looked out on
us in bewilderment. In our search we had strayed

over onto the roof of the Brazilian Legation. It seemed to cause him some surprise to see us doing second-story work on their house. It was a funny situation—but ended in another laugh. It is a good thing we can work in a laugh now and then.

The day was chiefly occupied with perfecting arrangements for getting off our German refugees. The Minister wished the job on me, and I with some elements of executive ability myself gave the worst part of it to Nasmith, the Vice-Consul-General. Modifications became necessary every few minutes, and Leval and I were running around like stricken deer all day, seeing the disheartening number of government officials who were concerned, having changes made and asking for additional trains. During the afternoon more and more Germans came pouring into the Consulate for refuge, until there were over two thousand of them there, terribly crowded and unhappy. Several convents were also packed, and we calculated that we should have two or three thousand to get out of the country. In the morning the Legation was besieged by numbers of poor people who did not know which way to turn and came to us because they had been told that we would take care of them. We were all kept busy; and Leval, smothering his natural feelings, came out of his own accord and talked and advised and calmed the frightened people in their own language. None of us would have asked him to do it, but he was fine enough to want to help and to do it without any fuss.

A crowd of curious people gathered outside the Legation to watch the callers, and now and then they boo-ed a German. I looked out of the window in time to see somebody in the crowd strike at a poor little worm of a man who had just gone out the door. He was excited and foolish enough to reach toward his hip pocket as though for a revolver. In an instant the crowd fell on him; and although Gustave, the messenger, and I rushed out we were just in time to pull him inside and slam the door before they had a chance to polish him off. Gustave nearly had his clothes torn off in the scrimmage, but stuck to his job. An inspired idiot of an American tourist who was inside tried to get the door open and address the crowd in good American, and I had to handle him most undiplomatically to keep him from getting us all into trouble. The crowd thumped on the door a little in imitation of a mob scene, and the Garde Civique had to be summoned on the run from the German Legation to drive them back and establish some semblance of order. Then de Leval and I went out and talked to the crowd—that is to say, we went out and he talked to the crowd. He told them very reasonably that they were doing harm to Belgium, as actions of this sort might bring reprisals which would cost the country dear, and that they must control their feelings. He sounded the right note so successfully that the crowd broke up with a cheer.

Orders have been issued to permit us free use of the telephone and telegraph, although they have been cut for everybody else. Yesterday afternoon I talked with the Consulates at Ghent and Antwerp. They were both having their troubles with Germans who wanted to get out of the country. I told them to send everybody up here and let them report at their own consulate, where they will be looked after.

The Government is taking no chances of having trouble because of the doings of francs-tireurs. The Minister of the Interior sent out, on the 4th, a circular to every one of the 2,700 communes in the country to be posted everywhere. The circular points out in simple and emphatic terms the duty of civilians to refrain from hostile acts and makes it clear that civilians might be executed for such acts. Aside from this, every newspaper in the country has printed the following notice signed by the Minister of the Interior:

TO CIVILIANS

The Minister of the Interior advises civilians, in case the enemy should show himself in their district:

Not to fight;

To utter no insulting or threatening words;

To remain within their houses and close the windows, so that it will be impossible to allege that there has been any provocation;

To evacuate any houses or small village which may be occupied by soldiers in order to defend themselves, so that it cannot be alleged that civilians have fired;

An act of violence committed by a single civilian would be a crime for which the law provides arrest and punishment. It is all the more reprehensible in that it might serve as a pretext for measures of repression resulting in bloodshed and pillage or the massacre of the innocent population with women and children.

In the course of the afternoon we got our telegrams telling of the appropriation by Congress of two and a half millions for the relief of Americans in Europe, and the despatch of the *Tennessee* with the money on board. Now all hands want some of the money and a cabin on the *Tennessee* to go home in.

——, the Wheat King, came into the Legation this morning and was very grateful because we contrived to cash out of our own pockets a twenty-dollar express check for him. He was flat broke with his pocket bulging with checks and was living in a *pension* at six francs a day. There is going to be a lot of discomfort and suffering unless some money is made available pretty soon. The worst of it is that this is the height of the tourist season and Europe is full of school-teachers and other people who came over for short trips with meager resources carefully calculated to get them through their traveling and home again by a certain date. If they are kept long they are going to be in a bad way. One of our American colony here, Heineman, had a goodly store of currency and had placed it at the disposal of the Legation, to be used in cashing at face value travelers' checks and other similar paper which bankers will not touch now with a pair of tongs. Shaler has taken charge of that end of the business and has all the customers he can handle. Heineman will have to bide his time to get any money back on all his collection of paper, and his contribution has meant a lot to people who will never know who helped them.

There was a meeting of the diplomatic corps last night to discuss the question of moving with the Court to Antwerp in certain eventualities. It is not expected that the Government will move unless and until the Germans get through Liège and close enough to

Her Majesty, Elisabeth, Queen of the Belgians

Mr. Brand Whitlock, American Minister to Belgium

threaten Louvain, which is only a few miles out of
Brussels. There was no unanimous decision on the
subject, but if the Court goes, the Minister and I will
probably take turns going up, so as to keep in communication with the Government. There is not much
we can accomplish there, and we have so much to do
here that it will be hard for either of us to get away.
It appeals to some of the colleagues to take refuge
with a Court in distress, but I can see little attraction
in the idea of settling down inside the line of forts and
waiting for them to be pounded with heavy artillery.

Liège seems to be holding out still. The Belgians
have astonished everybody, themselves included. It
was generally believed even here that the most they
could do was to make a futile resistance and get
slaughtered in a foolhardy attempt to defend their
territory against invasion. They have, however, held
off a powerful German attack for three or four days.
It is altogether marvelous. All papers have the head
lines: *"Les forts tiennent toujours."*

In the course of the afternoon we arranged definitely
that at three o'clock this morning there should be
ample train accommodations ready at the Gare du
Nord to get our Germans out of the country. Nasmith
and I are to go down and observe the entire proceedings, so that we can give an authoritative report afterward.

There is a German-American girl married to a
German who lives across the street from me. I sent
her word to-day that she and her husband and little
boy had better get away while there was a way open.
Last evening while we were at dinner at the Legation

the three of them arrived in a panic. They had heard that there was a mob of ten thousand people about the German Consulate about to break in and kill every German in the place. Of course they could not be persuaded to go near the Consulate or any of the other refuges. They wanted to settle down and stay at the Legation. As the Minister was on his way out to the meeting of the corps, the woman waylaid him, had got down on her knees and kissed his hand and groveled and had hysterics. He called for me and we got them quieted down. I finally agreed to go down to the Consulate and take a look so as to reassure them.

When I got there I found that the streets had been barred off by the military for two blocks in every direction, and that there was only a small crowd gathered to see what might happen. About as hostile as a lot of children. I got through the line of troops and in front of the Consulate found several hundreds of the refugees who had been brought out to be marched to the Cirque Royale, where they could be more comfortably lodged until it was time to start for the train. They were surrounded by placid Gardes Civiques and were all frightened to death. They had had nothing to do for days but talk over the terrible fate that awaited them if the bloodthirsty population of Brussels ever got at them; the stories had grown so that the crowd had hypnotised itself and was ready to credit any yarn. The authorities showed the greatest consideration they could under their orders. They got the crowd started and soon had them stowed away inside the Cirque Royale, an indoor circus near the Consulate. Once they got inside, a lot of them gave

way to their feelings and began to weep and wail in
a way that bade fair to set off the entire crowd. One
of the officers came out to where I was and begged me
to come in and try my hand at quieting them. I
climbed up on a trunk and delivered an eloquent
address to the effect that nobody had any designs on
them; that the whole interest of the Belgian Government lay in getting them safely across the frontier;
called their attention to the way the Garde Civique
was working to make them comfortable, and to reassure
them, promised that I would go with them to the
station, put them on their trains, and see them safely
off for the frontier. That particular crowd cheered up
somewhat, but I could not get near enough to be heard
by the entire outfit at one time, so one of the officers
dragged me around from one part of the building to
another until I had harangued the entire crowd on
the instalment plan. They all knew that we were
charged with their interests, and there was nearly
a riot when I wanted to leave. They expected me to
stay right there until they were taken away.

I came back to the Legation and told my people
that the way was clear and that they had nothing to
worry about. Mrs. Whitlock and Miss Larner had
taken the family in hand, were petting the baby boy,
and had them all cheered up to a sensible state of
mind. I got them into the motor and whisked them
down to the lines that were drawn about the block.
Here we were stopped and, sooner than undertake
a joint debate with the sentry, I was for descending
and going the rest of the way on foot. When a few
of the idly curious gathered about the car, the woman

nearly had a fit and scrambled back into the car almost in spasms. Of course the scene drew some more people and we soon had a considerable crowd. I gathered up the boy—who was a beauty and not at all afraid—and took him out of the car. There was in the front rank an enormous Belgian with a fiercely bristling beard. He looked like a sane sort, so I said to him: *"Expliquez à ces gens que vous n'êtes pas des ogres pour croquer les enfants."* He growled out affably: *"Mais non, on ne mange pas les enfants, ni leurs mères,"* and gathered up the baby and passed him about for the others to look at. My passengers then decided that they were not in such mortal danger and consented to get out. An officer I knew came along and offered to escort them inside. On the way in I ran into Madame Carton de Wiart, wife of the Minister of Justice, who was there to do what she could to make things run smoothly. She is rabid about the Germans, but is not for taking it out on these helpless people. And that seems to be the spirit of everybody, although it would be quite understandable if they showed these people some of their resentment. The Gardes were bestirring themselves to look after their charges. Some of them had contributed their pocket money and had bought chocolate and milk for the children and mineral waters and other odds and ends for those that needed them. And some of them are not very sure as to how long they will have pocket money for themselves. Aside from the fright and the heat and the noise of that crowd in the Cirque, it was all pretty depressing. During the night one old man died—probably from fright and shock—and a child was born. It was altogether a night of horror that could perfectly well have been avoided if people had only been able to keep calm and stay at home until time for the train to leave.

Having settled my charges and taken a look round, I went back to the Legation and got off some telegrams and talked with Bell over the telephone. He had a lot of news that we had not received and many errands to be done for people who had friends and relatives here.

A little after midnight friend Nasmith came along and we set out together for our rounds. We first took a look at one or two places and then went to my diggings for a sandwich and such rest as we could get before time to start on our round-up. Soon after midnight, Fortescue came rolling up in a cab looking for a place to lay his head. He had just come in from Liège, where he had had a close view of yesterday morning's heavy fighting. He said the Germans were pouring men in between the forts in solid formation, and that these sheep were being mown down by the Belgians heavily intrenched between the forts. The Germans are apparently determined to get some of their men through between the forts and are willing to pay the price, whatever it may be. To-day we hear that the Germans have asked for an armistice of twenty-four hours to bury their dead.

After we had hung upon his words as long as he could keep going, Nasmith and I got under way to look after our exodus. The Garde was keeping order at all places where there were refugees, and I was easy in my mind about that; my only worry was as to what might happen when we got our people out into the streets. Promptly at three o'clock we began to march them out of the Cirque. The hour was carefully chosen as the one when there were the least possible people in the streets; the evening crowds would have gone home and the early market crowd would hardly have arrived. A heavy guard was thrown around the people as they came out of the building and they were marched quickly and quietly down back streets to the Gare du Nord. I never saw such a body of people handled so quickly and yet without confusion. In the station four trains were drawn up side by side; as the stream of people began pouring into the station, it was directed to the first platform and the train was filled in a few minutes. At just the right moment the stream was deflected to the next platform, and so on until all four trains were filled. After starting the crowd into the station and seeing that there was going to be no trouble, I set off with an officer of the Garde Civique to see about other parties coming from some of the convents. They had not waited for us, but were already moving, so that when we got back to the station they tacked onto the end of the first party and kept the stream flowing.

As fast as the trains were filled, the signal was given and they pulled out silently. I stood behind some of the Garde Civique and watched the crowd pour in. The Gardes did not know who I was aside from the fact that my presence seemed to be countenanced by their officers, and so I overheard what they had to say. They were a decent lot and kept saying: *Mais c'est malheureux tout de même! Regardez donc ces pauvres gens. Ce n'est pas de leur faute,* and a lot more of that sort of thing.

It takes a pretty fine spirit to be able to treat the enemy that way. A lot of people in the passing crowd spotted me and stopped to say good-bye or called out as they went by. It was pathetic to see how grateful they were for the least kind word. I never saw such a pitiful crowd in my life and hope I never may again. They hurried along, looking furtively to right and left with the look of a rat that is in fear of his life. I have seldom pitied people more, for that sort of fear must be the most frightful there is—simple fear of physical violence.

It was remarkable to see the different classes of people who were there. The Manager of a bank of Brussels had abandoned everything he owned and joined the crowd. There were several financiers of standing who felt obliged to flee with their families. And there were lots of servants who had lived here for years and were really Belgian in everything but birth. Just before the last train left some closed wagons came from the prisons to bring a lot of Germans and wish them back on their own country in this way.

And there was not an incident. Here and there a prowling cab driver hooted, but there was not a stone thrown or any other violence. Before the last of the procession got into the station, it was nearly six o'clock and broad daylight. We moved up the platform with Major Dandoy and watched the last train leave. The Abbé Upmans was there through it all, working like a trump, bucking the people up; he did not stop until the last train pulled out into the fresh summer morning, and then he stayed aboard after the train was in motion to shake hands with a little handful of downhearted people. He shook himself and heaved a sigh of relief—remarking quietly that his duty had required him to go through all this and look after his charges while they were in trouble— but that now he might have the satisfaction of being a Belgian. I too heaved a sigh of relief, but it was because the mob was safely off and I need not worry about street fighting.

Dandoy had not had any sleep for nearly sixty hours, and though Nasmith and I were pretty tired ourselves, we thought the least we could do was to take him home. His family is in Liège and he has not been able to get any word from them. I offered to try a telephone message to the Consul at Liège, but have had no luck with it. None the less, Dandoy has been most grateful.

Before we left the station they began bringing in the wounded and prisoners. Most of the wounded I saw were not badly hurt, and were plucky and confident. Most of them were supported or led by Boy Scouts who have taken off the military the full burden of messenger work and a lot of other jobs. They are being of real value, as they can do lots of useful things and thereby release grown men for service at the front.

When I got back to the Rue St. Boniface—after stopping at the Legation to see what had come in— I had just time to throw myself down for a twenty-minute rest before the slave came in with my coffee. And then with no time for a tub, I had to hurry back and get into the harness. And none too soon, for the work began to pour in and I have been kept on the jump all day. If all goes well I hope to get to bed some time after midnight to-night. That means about three hours sleep and hard going during the past forty-eight hours.

This morning the various American committees came to the Legation to report on the measures they have taken for the protection of the colony in case of danger. I have been handed the pleasant task of Chief of Staff, with full authority to settle all matters affecting the protection of Americans in case hostilities reach this part of the country, as seems may well be the case before many days. In harmony with my well-known policy of passing the buck—more politely known as executive ability—I impressed Major Boyer of the Army, who is here for the time. He has set up an office at the headquarters of the committee and makes it his business to keep me fully posted as to what is going on there. First I started him out to look at the various houses that have been under discussion by the committee, so that he could decide as to their relative accessibility and general strategic advantages. He did this and made all sorts of arrangements tending to co-ordinate the work of the various sub-committees along the lines of the plan we drew up. It will be a great thing to have somebody who will act as buffer for all the detail and relieve me of just that much.

Germans who for one reason or another had not got away on our train kept turning up all day, and we kept sending them along to the Consulate. Late this afternoon the hard-working Nasmith came in to say that there were already seven hundred of them gathered there.   We shall have to have another special train for day after to-morrow morning, and hope to get most of the remaining Germans out of harm's way by that time.

The Belgians continue to be a surprise.   At last accounts they were still holding the forts at Liège. The French appear to have established themselves along the Meuse and to be ready for the attack when it comes.   Where the British troops are, nobody here seems to know—and, strange to say, they are not advertising their whereabouts.   There are plenty of people who have had confidential tips from their cook's brother, who lives in the country and has seen them with his own eyes.   According to such stories they are all landed at Ostend and are being hurried across the country through Malines.   Another story is that they have been shipped through to Liège in closed freight cars to outwit German spies, and that they are now in the thick of it.   According to still another of these confidential fellows, they have been shipped through Brussels itself in the night and we were unaware when they passed under our very windows.   You can choose any story you like and get an audience with it these days.

To-day's mouth-to-mouth news is that the French have fought a big battle near St. Hubert and repulsed the Germans with heavy losses.   This has about as much confirmation as the reports as to the whereabouts of the British army.

To-day trains have been coming in all day with wounded from Liège, and the lot—Belgian and German—are being cared for by the Red Cross.   The Palace has been turned into a hospital, and the Queen

has taken over the supervision of it. Nearly every big
hotel in town has turned its dining-room into a ward,
and guests are required to have their meals in their
rooms. Some of the big department stores have come
up finely in outfitting hospitals and workrooms, clearing out their stocks, and letting profits go hang for
the time being. The International Harvester Company
cleared its offices here and installed twenty-five
beds—informing the Red Cross that it would take care
of the running expenses as long as the war lasts. The
hospital facilities have grown far faster than the
wounded have come in, and there is an element of
humour in the rush of eager women who go to the
station and almost fight for the wounded as they are
brought off the trains.

I impressed the services of several people to help
out to-day, but the most valuable are two crack
stenographers who have been turned over to us by
business firms here. By dint of labouring with them
all morning and afternoon and seeing as few people as
possible, I have managed to clean up my desk, so that
I can go to bed with a clear conscience to-night when
I have got through my call to London.

Brussels, August 8, 1914.—To-day our new organisation is working like clockwork. In Cruger's formerly
calm chancery there are five typewriters pounding
away, and at the committee rooms there are swarms
of people working to take care of odds and ends.
Monsieur de Leval has a table at one side of my
room, and the committee relieves us of the people
who want information and those who want to talk.

Sunday, August 9th.—I got this far when the roof fell in last night. During the afternoon yesterday I got out to attend to a few odds and ends of errands— and, as always happens when I go out, things began to happen. I came back to find the Minister and de Leval wrestling with a big one.

A curious telegram had come from The Hague, quoting the text of a message which the German Government desired us to present to the Belgian Government. Here it is in translation, a truly German message:

The fortress of Liège has been taken by assault after a brave defense. The German Government most deeply regret that bloody encounters should have resulted from the attitude of the Belgian Government toward Germany. Germany is not coming as an enemy into Belgium; it is only through the force of circumstances that she has had, owing to the military measures of France, to take the grave decision of entering Belgium and occupying Liège as a base for her further military operations. Now that the Belgian army has upheld the honour of its arms by its heroic resistance to a very superior force, the German Government beg the King of the Belgians and the Belgian Government to spare Belgium further horrors of war. The German Government are ready for any compact with Belgium which can be reconciled with their conflicts with France. Germany once more gives her solemn assurance that it is not her intention to appropriate Belgian territory to herself and that such an intention is far from her thoughts. Germany is still ready to evacuate Belgium as soon as the state of war will allow her to do so.

Of course we were loath to present anything of the sort, but the thing had to be handled carefully. After some pow-wowing I went over to the Foreign Office with the message and saw Baron van der Elst. I told him seriously that we had received a very remarkable telegram which purported to contain a message from the German Government; that it bore no marks of authenticity, and that we were not sure as to its source; but that we felt that we should be lacking in frankness if we did not show him what we had received. He seized the message and read it through, his amazement and anger growing with each line. When he had finished, he gasped for a minute or two and then led me into the next room to the Minister for Foreign Affairs, M. Davignon, to whom he translated the telegram aloud. When they had finished discussing the message and I had a pretty clear idea as to the Belgian attitude toward the proposal—not that I had had any real doubt—I asked him: "If the American Minister had delivered this message what would have been its reception?" Without an instant's hesitation, M. Davignon replied: "We should have resented his action and should have declined to receive the communication."

That was all I wanted to know and I was ready to go back to the Legation.

I took Baron van der Elst home in the car and had the pleasure of seeing him explain who he was to several Gardes Civiques, who held up the car from time to time. He was very good-natured about it, and only resented the interruptions to what he was trying to say. His son is in the army and he has no news of him. As he got out of the car he remarked that if it were not so horrible, the mere interest of events would be enough to make these days wonderful.

When I got back to the Legation and reported the result of my visit, we went to work and framed a telegram to Washington, giving the text of the German message, explaining that we had nothing to prove its authenticity and adding that we had reason to believe that the Belgian Government would not accept it. The same message was sent to The Hague. This pleasant exercise with the code kept us going until four in the morning. Eugène, the wonder chauffeur, had no orders, but curled up on the front seat of his car and waited to take me home. He was also on hand when I got up a couple of hours later, to take me back to the Legation. Chauffeurs like that are worth having.

When I came in this morning the place was packed with Germans. Some cheerful idiot had inserted a notice in the papers that all Germans were to be run out of the country, and that they should immediately apply to the American Legation. As the flood poured in, Leval got on the telephone to the Sûreté Publique and found out the true facts. Then we posted a notice in the hall. But that was not enough. As is always the case with humans, they all knew better than to pay any attention to what the notice said and each one of the hundred or more callers had some reason to insist on talking it over with somebody. When they once got hold of one of us, it was next to impossible to get away without listening to the whole story of their lives. All they had to do was to go down to the German Consulate-General, where we had people waiting to tell them all there was to know. It was hard to make them realise that by taking up all our time in this way, they were preventing us from doing things that were really necessary to serve them in more important matters. I said as much to several of them, who were unusually long-winded, but every last one replied that HIS case was different and that he must be heard out at length.

Our refugee train left this morning and took eight hundred more of the poor people. Where they all turn up from, I don't know, but each day brings us a fresh and unexpected batch. Many of the cases are very sad, but if we stop to give sympathy in every deserving case, we should never get anything practical done for them.

To-day's budget of news is that the French have got to Mulhouse and have inflicted a decisive defeat upon the Germans. According to reports, the Alsatians went mad when the French troops crossed the frontier for the first time in forty-four years. They tore up and burned the frontier posts and generally gave way to transports of joy. I would have given a lot to see the crowds in Paris.

A letter came yesterday from Omer, the legation footman, who is at Tirlemont with the artillery. He said he had not yet been hit, although he had heard the bullets uncomfortably near. He wound up by saying that he had *beaucoup de courage*—and I believe him.

It seems that some of the German troops did not know what they were attacking and thought they were in France. When brought here as prisoners, some of them expressed surprise to find that Paris was so small. They seem to have thought that they were in France and the goal not far away.

The King to-day received through other channels the message from the Emperor of Germany in regard

to peace, which we declined to transmit. I have not
seen its text, but hear it is practically identical with
the message sent us, asking the King to name his
conditions for the evacuation of Liège and the abandon-
ment of his allies, so that Germany may be entirely
free of Belgian opposition in her further operations
against France. I have heard among Belgians only
the most indignant comments on the proposal and
look forward with interest to seeing the answer of the
King, which should appear to-morrow.*

The town is most warlike in appearance. There
is hardly a house in the town that does not display
a large Belgian flag. It looks as though it were be-
decked for a fiesta. Here and there are French and
British flags, but practically no others. Every motor
in town flies a flag or flags at the bow. We fly our
own, but none the less, the sentries, who are stationed
at all the corners dividing the chief quarters of the
town and before all the Ministries and other public
buildings, stop us and demand the papers of the
chauffeur and each passenger in the car. We have
passports and all sorts of other papers, but that was
not enough, and we finally had to be furnished by

---

* The Belgian reply, which was sent on August 12th through
the Netherlands Minister for Foreign Affairs, is as follows:

The proposal made to us by the German Government repeats
the proposal which was formulated in the ultimatum of August
2nd. Faithful to her international obligations, Belgium can only
reiterate her reply to that ultimatum, the more so as since August
3rd, her neutrality has been violated, a distressing war has been
waged-on her territory, and the guarantors of her neutrality have
responded loyally and without delay to her appeal.

the Ministry for Foreign Affairs with a special *laisser-passer*. This afternoon I slipped out for a breath of air and was held up and told that even that was no good until I had had it viséd by the military authorities. It is said that these strict measures are the result of the discovery of a tremendous spy system here. According to the stories which are told, but of which we have little confirmation, spies are being picked up all the time in the strangest disguises.

The gossip and "inside news" that is imparted to us is screamingly funny—some of it.

Yesterday, according to one of these yarns, four nuns arriving at the Gare du Midi were followed for some time and finally arrested. When searched, they proved to be young German officers who had adopted that dress in order to conceal carrier pigeons which they were about to deliver in Brussels. Wireless outfits are said to have been discovered in several houses belonging to Germans. I cannot remember all the yarns that are going about, but even if a part of them are true, it should make interesting work for those who are looking for the spies. The regular arrests of proven spies have been numerous enough to turn every Belgian into an amateur spy-catcher. Yesterday afternoon Burgomaster Max was chased for several blocks because somebody raised a cry of *"Espion"* based on nothing more than his blond beard and chubby face. I am just as glad not to be fat and blond these days.

Yesterday afternoon a Garde Civique came in with the announcement that the chancellor and clerks of the German Legation, who were locked up there, were

in dire distress; that a baby had been born the day
before to the wife of the concièrge, and that all sorts
of troubles had come upon them. Leval, who had
announced that his heart was infinitely hardened
against all Germans, was almost overcome by the
news of a suffering baby and ran like a lamp-lighter
to get around there and help out. When we arrived,
however, we found them all beaming and happy. The
baby had been born some days before and the mother
was up and about before the Legation had been closed.
Their meals are sent in from a neighbouring restaurant,
and they are perfectly contented to bide their time
as they are. They had orders from Berlin not to leave
the Legation, so it made little difference to them
whether they were blockaded by the Belgian authorities
or not. I shall drop in every day or two and see
whether there is anything I can do to lighten their
gloom. Of course their telephone was cut off and they
are not allowed to receive mail or papers, so they
are consumed with curiosity about developments. It
was, of course, necessary to refuse to answer their
questions about what was going on and to make
assurance doubly sure, I had the Garde Civique stand
by me while I talked with them.

As things shape up now it looks as though we were
the only life-sized country that could keep neutral
for long, and as a consequence all the representatives
of the countries in conflict are keeping us pretty well
posted in the belief that they may have to turn their
interests over to us. We shall probably soon have to
add Austrian interests to the German burdens we now
have. If there is a German advance, some of the

Allied ministers will no doubt turn their legations over
to us. The consequence is that we may see more of
the inside of things than anybody else. Now, at least,
we are everybody's friends. This is undoubtedly the
most interesting post in Europe for the time being,
and I would not be anywhere else for the wealth of the
Indies.

Brussels, Aug. 10, 1914.—The Belgian Government has finally got out a proclamation, urging German subjects to leave the country, but stating that
in the event of a general order of expulsion, certain
classes of people will be allowed to remain, such as,
very old persons, the sick, governesses, nurses, etc.,
and even others for whom Belgians of undoubted
reputation are willing to vouch. There are quantities of Germans who have lived here all their lives,
who are really more Belgian than German, have
no interest in the present conflict and are threatened
with financial ruin if they leave their interests here,
and it is pretty hard on them if they are to be obliged
to get out, but they are only a few of the many, many
thousands who are suffering indirectly from the effects
of the war. It is not any easier for the manufacturers
in the neighbourhood of Liège, who will see the work
of many years wiped out by the present hostilities.
Some inspired idiot inserted in the papers yesterday
the news that the Legation was attending to the
repatriation of German subjects and the consequence
is that our hallways have been jammed with Germans
all day, making uncouth noises and trying to argue
with us as to whether or not we are in charge of Ger-

man interests. The mere fact that we deny it is not enough for them! I suppose that the hallways will continue to sound like a celebration of Kaisersgeburts-tag until we have sent off the last of them.

This morning a large, badly frightened darkey came in looking for a passport. He awaited his turn very quietly, and grew visibly more and more apprehensive at the long series of questions asked of the people ahead of him. When he moved up to the desk, the first question was:

"Where do you want to go?"

"Jes as fur as the stature of Libbuty."

"Are you an American citizen?"

"Me? Lawd bless yuh! No, I ain't nuthin' but a plain ole Baltimoh coon."

Then they gave him the usual blank to fill out. One of the questions on it was:

"Why do you desire to return to the United States?"

Without any hesitation he wrote:

"I am very much interested in my home at the present time."

Everybody here is intensely curious as to what has become of the British army; the most generally accepted story is that troops have been landed at Calais, Dunkirk and Ostend, but although this is generally believed, there seems to be absolutely no official confirmation of it. Everyone seems to take it for granted that the British will turn up in good form when the right time comes, and that when they do turn up, it will have a good effect. If they can get to the scene of hostilities without everybody knowing about it, it increases by just so much their chances

Mr. Brand Whitlock, American Minister to Belgium. Taken during a Fourth of July luncheon at the Royal Golf Club

Burgomaster Max

of success and anyone that knows anything at all is
keeping mum and hoping that no British soldier will
stumble over a chair and make a noise and give away
the line of march.

Our letters from London indicate intense satisfaction
with the appointment of Kitchener and confidence
that he will get a maximum of service out of the forces
at his command.

We have been looking from one moment to another
for news of a big naval engagement, but suppose the
British Navy is somewhere waiting for a chance to
strike.

Colonel Fairholme, the British Military Attaché,
has made a number of trips to the front and reports
that the morale of the Belgian troops is excellent, that
the organisation is moving like clockwork, and, as he
expresses it, that "every man has his tail up."

This evening I went over to the British Legation
to see the Colonel, and learn whatever news he had
that he could give me. There was a great scur-
rying of servants and the porter was not to be found
in the chancery. The door to Grant-Watson's room
was ajar, so I tapped, and, on being bade in a
gruff voice to "Come in," walked into the presence
of a British officer in field uniform, writing at Webber's
desk. He was dusty and unshaven, and had evidently
come in from a long ride. I promptly backed out
with apologies and was hustled out of the place by
Kidston, who came running out from the Minister's
office. I asked him if the rest of the army was hidden
about the chancery, and his only reply was to tell me
to run along and find the navy, which they themselves

had not been able to locate. They evidently have all they need to know about the whereabouts of the army, but have succeeded in keeping it dark.

C. M. came over to the Legation this afternoon to get some books for her mother. We fixed her up and put her in her car, when she announced that on the way over she had been arrested and taken to the police station as a German. People are pointing out spies on the street, and anybody that is blond and rosy-cheeked stands a fine show of being arrested every time he goes out. She had impressed this car with a suspected number and paid for it by being made into a jail bird.

My day's work began with a visit to the German Legation. The Government asked me to secure and return the number for the automobile of von Stumm, the German Counselor. I had his machine put in the Legation the day after he left, although he had offered it to me. I presented myself at the door of the Legation with the note from the Foreign Office, asking for the number, but was refused admittance by the Gardes Civiques. They were very nice, but stated that they had the strictest orders not to let anybody come in or out, and that they had not discretionary powers. At a visit at the Foreign Office later in the day, I told of my experience and asked that I be furnished by the military authorities with a *laisser-passer* which would enable me to enter the Legation whenever I so desire. This afternoon I received a formidable document from the Military Governor which gives me free passage—so far as I can make out—to enter the Legation in any way save by telephone or telegraph.

I shall go around to-morrow and rub it in on the
Gardes Civiques.

The question of passes has been changed and made
more strict each day, and has got to be a sort of joke.
I first used my card, that was declared insufficient
almost from the first. Then I tried my *permis de
circulation*, which was issued to allow me to get into
the railway stations without paying. That was good
for a day or so. Then I tried my passport (as a bearer
of despatches), and that got me through once or
twice. Then the Minister for Foreign Affairs gave me
his personal card with a *laisser-passer* in his own hand,
but that was soon turned down on the ground that
the military authorities are in control and the civil
authorities cannot grant passes. Finally the Government has got out a special form of *laisser-passer* for
the diplomats, and it may prove to be good—although
it is not signed by the military authorities. I have
taken the precaution of keeping all the aforementioned
documents and some others on my person, and am
curious to see how soon I shall have to have some
other. The Garde Civique is no longer content with
holding up the car every few blocks and examining
the *pièce d'dentité* of the chauffeur; they must now be
satisfied as to the bona fides of each passenger. Doing
some errands around town this afternoon I was held
up and looked over eleven times. I now pull out all
the documents I own and hand out the bunch each
time I am stopped. The Garde then, in most cases,
treats the matter rather humorously, and the next
time I pass lets me go on without going through the
whole performance again. In front of the German

Legation, however, which we nearly always pass on our way to or from town, we are invariably held up and looked into seriously. I know most of the people on the different shifts by this time and wish them well each time they look at the well-remembered papers. I shall keep the credentials and any others that may eventually be added to them, and perhaps some day I shall be able to paper a room with them.

In the course of the morning there were several matters of interest which made it necessary for me to go to the Foreign Office. All their messengers are now gone, and in their place there is a squad of Boy Scouts on duty. I had a long conference with van der Elst, the Director-General of the Ministry. In the course of our pow-wow it was necessary to send out communications to various people and despatch instructions in regard to several small matters. Each time van der Elst would ring, for what he calls a "scoots," and hand him the message with specific instructions as to just how it should be handled. The boys were right on their toes, and take great pride in the responsibility that is given them. Some of them have bicycles and do the messenger work through the town. Those who have not, run errands in the different buildings and attend to small odd jobs.

The Red Cross is very much in evidence. I went around to the headquarters after my call at the Foreign Office, to make a little contribution of my own and to leave others for members of our official family. The headquarters is at the house of Count Jean de Mérode, the Grand Marshal of the Court. The entrance hall was filled with little tables where women

sat receiving contributions of money and supplies.
I had to wait some time before I could get near enough
to one of the dozen or more tables, to hand in my
contributions. This is the headquarters, but there are
any number of branch offices, and they are said to be
equally busy. The society has been quite overcome
by the way people have come forward with gifts, and
they have been almost unable to get enough people
together to handle them as they come in. The big
cafés down-town nearly all have signs out, announcing
that on a certain day or days they will give their entire
receipts to the Red Cross or to one of the several
funds gotten up to take care of those suffering directly
or indirectly from the war. Many of the small shops
have signs out of the same sort, announcing that the
entire receipts for all articles sold on a certain day
will be handed to one of the funds. They must have
gathered an enormous amount of money, and I don't
doubt they will need it. The wounded are being
brought in in great numbers and many buildings are
quite filled with them. In nearly every street there
is a Red Cross flag or two, to indicate a temporary
hospital in a private house or a hotel or shop, and
people are stationed in the street to make motors
turn aside or slow down. There are almost no motors
on the street except those on official business or Red
Cross work; and, because of the small amount of
traffic, these few go like young cyclones, keeping
their sirens going all the time. The chauffeurs love
it and swell around as much as they are allowed to
do. I pray with ours now and then, but even when
I go out to the barber, he seems to believe that he

is on his way to a fire and cuts loose for all he is worth.

Quantities of German prisoners continue to be brought here for safe keeping, and many of them are taken on down to Bruges. Among those removed there for unusually safe keeping yesterday was a nephew of the Emperor.

Judging from the stories printed in the *London Times* which arrived to-night, the German Government aroused great enthusiasm by playing up the capture of Liège. The Germans evidently were led to believe they had gained a great victory; whereas the forts, which are the only object of the campaign, are still intact. The city itself is undefended, and there is no great military reason why the Belgians should not allow it to be taken. The German troops that had invested the town have not taken over the administration, but appear to be confining themselves to requisitioning provisions and supplies, of which they are in need. The Berlin papers made a great hurrah about the capture of the citadel, which is a purely ornamental old fort without military importance. From what they tell me, I judge that you could back an American army mule up against it and have him kick it down without the expense of bombarding it. It sounds well in the despatches, however.

Eight French aeroplanes sailed over the city this afternoon, probably coming from Namur. One of the machines landed on the aviation field at the edge of the city, and the aviator was nearly torn to shreds by admirers who wanted to shake him by the hand and convince him that he was really welcome to

Brussels. It is said that some of these fellows are going to lie in wait for the Zeppelins which have been sailing over Brussels by night to terrify the population. We hear that one of the Belgian army aviators did attack a Zeppelin and put it out of business, bringing to earth and killing all the crew. He himself went to certain death in the attempt.

The afternoon papers say that in Paris the name of the Rue de Berlin has been changed to Rue de Liège. Here the Rue d'Allemagne has been changed to Rue de Liège and the Rue de Prusse to Rue du Général Leman, the defender of Liège. The time abounds in *beaux gestes* and they certainly have their effect on the situation.

Kitchener says that the war may last for some time. At first it seemed to be taken for granted that it could not last long, as the financial strain would be too great and the damage done so enormous that one side or the other would have to yield to avoid national bankruptcy.

Brussels, August 11, 1914.—Our halls have been filled with Germans and Americans, the latter in smaller numbers and the former in larger crowds than ever. They are gradually being got out of the country, however, and those who are going to remain are being induced to go to the right authorities, so that their troubles will soon be settled to a large extent, and they will not be coming here so much. We are getting off hundreds of telegrams about the whereabouts and welfare of Americans and others here and in other parts of Europe; this work alone is enough to keep a good-sized staff working, and we have them hard at it.

This afternoon I went over to the British Legation and saw Colonel Fairholme, the military attaché, for a few minutes. He was just back from a trip out into the wilds with a party of British officers and was so clearly rushed that I had not the heart to detain him, although I was bursting with curiosity about the news he evidently had concealed about him. He appreciates the lenient way I have treated him, and goes out of his way to let me have anything that he can.

While I was out we saw a German monoplane which sailed over the city not very high up. The newspapers have published a clear description of the various aeroplanes that are engaged in the present war, so that nobody will be foolish enough to fire at those of the allies when they come our way. This one was clearly German, and the Garde Civique and others were firing at it with their rifles, but without any success. Our Legation guard, which consists of about twenty-five men, banged away in a perfect fusillade, but the airman was far too high for them to have much chance of hitting him.

Yesterday afternoon when the German biplanes passed over the city, a Belgian officer gave chase in a monoplane, but could not catch them. Contests of this sort are more exciting to the crowd than any fancy aviation stunts that are done at exhibitions, and the whole town turns out whenever an aeroplane is sighted.

This morning I presented myself at the German Legation with the imposing *laisser-passer* furnished

Belgian War Medals

Belgian War Medals

me by the Military Governor of Brabant, but the
guard on duty at the door had not received orders to
let me in and turned me down politely but definitely.
I took the matter up with the Foreign Office and said
that I wanted it settled, so that I would not have any
more fruitless trips over there. At five an officer from
the État-Major of the Garde Civique came for me in
a motor and took me over to the Legation, to give
orders in my presence that whenever I appeared I was
to be allowed to pass without argument. As I got into
the motor I noticed that the soldier who was driving
the car looked at me with a twinkle in his eye, but
paid no attention to him. When I took a second look
I saw that it was G. B——, with whom I had played
golf several times. I am constantly being greeted by
people in uniform whom I had known at one time or
another. It is hard to recognise them in uniform.

So far as operations in Belgium are concerned, we
may not have anything big for some days to come;
but, in the meantime, work of preparation is being
pushed rapidly and supplies and reinforcements are
being rushed to the front. Half the shops in town are
closed, and all the people are working either in the
field or taking care of the wounded or prisoners. There
are said to be some eight thousand German prisoners
in Belgium, and it is some work to take care of them
all.

Brussels, August 12, 1914.—A few minutes' gap, so
I seize my pen to scratch off a line.

Last night when I left here I rode up the Rue
Bélliard on my way home. I was stopped in front

of the German Legation by the guard which was placed across the street. They examined the chauffeur's papers carefully and then looked over mine. They compared the tintype on my *laisser-passer* with the classic lineaments of the original, and after looking wise, told me to move on. When we got up to the Boulevard there was great cheering, and we came out on a thin file of French cavalry, which was on its way through town from the Gare du Midi. The crowd was mad with enthusiasm and the soldiers, although plainly very tired, pulled their strength together every now and then to cry, "*Vive la Belgique!*" There were crowds on the Boulevards, waiting for news from *lá bas*. A few French officers were going about in cabs, and each time that one appeared the crowd went mad. The officers were smiling and saluting, and every now and then one stood up in his place and cheered for Belgium. In twenty minutes or so, I saw that we could get through, so started for home and bed.

When we got to the Porte de Namur, we heard frenzied cheering down by the Porte Louise. The chauffeur is a regular old war horse who does not want to miss a trick. He cast a questioning glance over his shoulder; and, catching my nod, put on full speed down the Boulevard until we came to a solid crowd banked along the line of march of more French cavalry. The people in the crowd had bought out the nearby shops of cigars and cigarettes and chocolate and small flasks of brandy, and as each man rode by, he was loaded up with as much as he could carry. The défilé had been going on for over an hour, but the enthusiasm

was still boundless. All the cafés around the Porte
Louise sent out waiters and waitresses with trays of
beer to meet the troops as they came into the Avenue
Louise. Each man would snatch a glass of beer,
swallow it as he rode along and hand it back to others
who were waiting with empty trays a hundred yards
or so down the line of march. The men were evidently
very tired, and it was an effort for them to show any
appreciation of their reception, but they made the
effort and croaked out, *"Vive la Belgique!"* The
French and British troops can have anything they want
in this country. They will be lucky, though, if they
escape without acute indigestion.

Yesterday afternoon, as I was coming out of the
chancery of the British Legation, a little cockney
messenger in uniform came snorting into the court
on a motor-cycle. As he got off he began describing
his experiences, and wound up his story of triumphant
progress—"And when I got to the Boulevards I ran
down a blighter on a bicycle and the crowd gave me
an ovation!"

More troubles to-day about the German Legation.
The État-Major gave orders that nobody but I
should be allowed to enter. The laymen who have
the onerous duty of protecting the Legation held
a council of war, and decided that this precluded
them from allowing food to go in; so when the waitress
from the Grand Veneur with the lunch of the crowd
inside came along, she was turned back and told I
should have to go with her. I went around to the
Legation and fixed it up with the guard. A few minutes
ago the waitress came back with word that more

bread and butter was wanted, but that the guard had changed and that she was again barred out. Monsieur de Leval and I went around again and fortunately found some one from the État-Major who was there for inspection. He promised to get proper orders issued and now we hope that we shall not be obliged to take in every bite under convoy.

There are ominous reports to-day of a tremendous German advance in this direction, and it is generally believed that there will be a big engagement soon near Haelen, which is on the way from Liège to Tirlemont. Communications are cut, so I don't quite see where all the news comes from.

After dinner.—News sounds better to-night. Although there is nothing very definite, the impression is that the Belgians have come out victorious to-day in an engagement near Tirlemont. I hope to get some news later in the evening.

During a lull in the proceedings this afternoon, I got in Blount's car and went out to Brooks, to see his horses and arrange to have him send them in for our use every afternoon. He came over here a few months ago to spend the rest of his life in peace and quiet. It looks as though he wouldn't get much of either.

The Avenue de Tervueren, a broad boulevard with a parkway down the centre, is the most direct way into town from the scene of the fighting, and there has been a general belief that the Germans might rush a force into town in motors that way. In order to be ready for anything of the sort, a barricade has been made of heavy tram cars placed at right angles across the road, so that they do not absolutely stop traffic,

The Marquis de Villalobar, Spanish Minister at Brussels

A barbed wire entanglement at Antwerp

The Garde Civique's idea of a barbed wire entanglement at the beginning of the war.   (Taken at the end of the Avenue Louise)

but compel motors to slow down and pick their
way, thus:

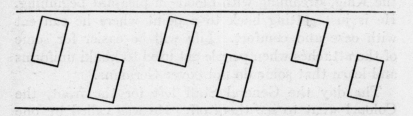

It is close work getting through, and can only be
done at a snail's pace.

The latest news we have is that the nearest large
German force is just 38 miles away from Brussels.

Brussels, August 13, 1914.—Last night, after dining
late, I went out to find my friend, Colonel Fairholme,
and see if he had any news. He had just finished his
day's work and wanted some air. Fortunately I had
the car along and so took him out for a spin to the
end of the Avenue Louise. We walked back, followed
by the car, and had a nightcap at the Porte de Namur.

The Colonel has been going to Louvain every day,
to visit the General Staff and report to the King as
the military representative of an ally. The first time
he arrived in a motor with Gen. de Selliers de Moran-
ville, the Chief of Staff. As they drew into the square
in front of the headquarters, they saw that everything
was in confusion and a crowd was gathered
to watch arrivals and departures. When their car
stopped, a large thug, mistaking him for a German
officer, reached in and dealt him a smashing blow on

the mouth with his fist, calling him a *"sal alboche"*
by way of good measure. He had to go in and report to
the King, streaming with blood—a pleasant beginning.
He is just getting back to a point where he can eat
with ease and comfort. Life will be easier for some
of the attachés when people get used to khaki uniforms
and learn that some do not cover Germans.

The day the General Staff left for the front, the
Colonel went to see them off. He was called by one
of the high officers who wanted to talk to him, and
was persuaded to get on the train and ride as far as the
Gare du Luxembourg, sending his car through town
to meet him there. Word came that the King wanted
to see the Chief of Staff, so he asked the Colonel to
take him to the Palace. When the crowd saw a British
officer in uniform and decorations come out of the
station accompanied by the Chief of Staff and two
aides, they decided that it was the Commander-in-
Chief of the British army who was arriving and gave
him a wonderful ovation. Even the papers published
it as authentic. He was tremendously fussed at the
idea of sailing under false colors, but the rest of us
have got some amusement out of it.

Stories are coming in here about the doings of the
German troops. According to reports they came into
Hasselt and took the money in the town treasury and
the local bank—some two and a half millions altogether.
The story, whether true or not, has caused a great
deal of ill feeling here. There is another story that
the commanding officer of one of the forts around
Liège was summoned to parley with a white flag.
When he climbed on top of his turret, he was shot

through both legs and only saved by his men pulling him to cover.   Of course there are always a great many stories of this sort scattered broadcast at the beginning of every war, but in this instance they seem to be generally believed and are doing the Germans no good at all.

Mlle. D——, one of our stenographers, has a brother in the French army.   She has not heard a word from him since the war began, and had no idea where he was.   Yesterday a small detachment of French cavalry came along the street.   She ran out, called to one of them that her brother was in the ——, and asked where it was.   They told her it had not yet been in action and she has been walking on air ever since. But she could not telegraph the good news to her family, for fear of betraying military movements.

Roger de Leval, the 8-year-old son of our friend, practically broke off diplomatic relations with his father and mother because he was not allowed to be a Boy Scout.   His father was at the Legation, his mother at the Red Cross, and he had to stay at home with his governess.   He felt so badly about it that we had Monsieur de Leval register him as a B. S., and have him assigned to special duty at the Legation. He attends in full uniform and carries messages and papers from my room to the other offices and vice versa.   When we go out he rides on the box with the chauffeur and salutes all the officers we pass. They are used to it now and return the salutes very gravely.   The youngster now feels that he is really doing something, but is outraged because we go along. He wants to undertake some of the big missions alone.

Princesse Charles de Ligne was in this morning. Her son, Prince Henri, head of that branch of the house, has enlisted as a private in the aviation corps. There seemed to be no way for him to have a commission at once, so he put his star of the Legion of Honor on his private's uniform and was off to the front yesterday. That's the spirit.

Comtesse d'A —— was at their home in the Grand Duchy when war broke out. No news had been received from her, and her husband was worried sick. We got a message through via The Hague and got word back this morning that she was safe and well. I went up to tell him the good news. He was presiding over some sort of committee meeting, and the maid said I could not see him. I insisted that she should announce me and after some argument she did. As the door opened, the buzz subsided and she announced: "*Monsieur le Secrétaire de la Légation d'Amérique.*" There was a terrible cry of fear and the old Count came running out white as a sheet. Before he had come in sight I called out, "*Les nouvelles sont bonnes!*" The old chap collapsed on my shoulder and cried like a baby, saying over and over: "*J'étais si inquiet: j'étais si inquiet!*" He soon pulled himself together and showed me out to the car with the honours of war. We send and receive hundreds of telegrams of inquiry and shoot them through in a perfectly routine way. It is only now and then that we come to a realising sense of the human side of it all.

This afternoon I went over and made inquiry as to the well-being of those who are cooped up in the German Legation. They are getting along perfectly well, but are consumed with curiosity as to the progress of the war. The Government has not allowed them to have any letters or newspapers, and they are completely in the dark as to what is going on. I felt like a brute to refuse them, but could not very well do anything against the wishes of the Government. They were decent enough not to embarrass me by insisting, which made it harder to refuse. The son of Hofrath Grabowsky, the Chancellor of the Legation, is Secretary of the German Consulate at Antwerp. He came down here to say good-bye to his father the day war was declared, and lingered so long that he was cooped up with the others. He is liable for military service in Germany, and having left his post at Antwerp at such a time, he must face a court martial whenever he does get home. There are five or six people there, including the wife of the old Hofrath, who are firmly convinced that they will all be murdered in their beds. It is my daily job to comfort them and assure them that nobody now here is giving any thought to them.

Last night I dined with Colonel Fairholme and Kidston, the First Secretary of the Legation. We went to the usually crowded terrace of the Palace Hotel, where we had no difficulty in getting a table in the best part of the balcony. The few other diners were nearly all colleagues or officers. Military motors and motorcycles came and went, and orderlies dashed up on horseback and delivered messages; it looked like war.

The proprietor of the hotel, who has given one hundred thousand francs to the Red Cross, rolled up in his motor from a trip to the front and got out with an armful of Prussian helmets and caps, which he had collected. A crowd gathered round the motor and displayed as much pleasure as though he had brought in a whole German Army corps. The novelty of these souvenirs has not yet worn off.

Women with big tin boxes came by every few minutes to collect for the Red Cross or some other fund. Finally the Colonel protested, and asked if there was no way of buying immunity. That was quickly arranged by giving up five francs, in return for which we were given tags of immunity. Dozens of collectors came by during the evening, but our ostentatiously displayed tags saved us.

We ate at our leisure—out of doors—the first unhurried and unharried meal I have had for days, and then got back to the Legation.

This afternoon the Minister and I went over to see Sir Francis Villiers, the British Minister, and spent half an hour with him. He is evidently all ready to make a quick get-away whenever it looks as though the Germans would come to Brussels. A number of the other diplomats are also prepared to depart. Those who are accredited at The Hague will probably go there, and the others will go to Antwerp. We are too busy here to enjoy the luxury of spending a month undergoing a siege, so no matter what happens, we shall probably not go along. The Minister and I shall take turns from time to time, going up to pay our respects.

Having some things to talk over, the Minister and I went for a drive after our visit, and it was well we did, for when we got back, we found the hall filled with callers. As the tourists and the Germans leave, the war correspondents begin to come in, and in a few days we shall probably have the place full of them. I heard to-day that there were 200 of them in London, and that most of them want to come on here.

Maxwell, the British correspondent, told me this afternoon that he looked for a big engagement at Diest to-morrow or the day after. He has been down through the fighting zone ever since the trouble began, and probably knows more about pending operations than any other civilian.

While I was writing, Z—— came in, suffering from a bad case of panic. He announced as he burst into my office that the Germans were within 20 kilometers of Brussels and were going to occupy the city this evening. He was fairly trembling, but got indignant because I denied it, having just talked with Colonel Fairholme and with Maxwell, both of whom had no more than come back from the front. The fact that it had been published in the *Soir* was enough for him, and although the news had made him nervous, he hated to have his perfectly good sensation spoiled.

The authorities, so as to be prepared for any eventuality, have this evening published a communiqué to impress upon the population the necessity for abstaining from any participation in the hostilities in case of an occupation. It advises everybody to stay indoors and avoid any words or actions that might give an excuse for measures against non-combatants.

August 15th.—Last night I dined with the Colonel, Grant-Watson, and Kidston at the Palace. I was looking forward to a lot of interesting talk, as the Colonel had just come from the front. Just as we were settling

down to our conversational Marathon, up walked
——, the —— Chargé and bade himself to dine
with us. He is strongly pro-German in his sympathies,
and, of course, that put a complete damper on conversation. We talked about everything on earth save
the one thing we were interested in, and sat tight in
the hope that he would move on. Not only did he
stay, but after a time the ——— First Secretary came
and joined us, and we gave up in despair. The only
result of the evening was that I gathered the impression that there is a good deal of apprehension on
the part of the allies as to the result of the next big
battle, which may occur any day now. The Germans
are undoubtedly pretty near now, perhaps a good deal
nearer than we know. Just before dinner the War
Office announced that there would be no further
official communiqués as to the operations. That looks
as though they were battening down the hatches for
the next big engagement.

Yesterday's papers announced France's declaration
of war against Austria. This morning comes the news
that Montenegro has also declared her intention of
wiping Austria off the map. Our daily query now is—
"Who has declared war to-day?"

Every minute we are not hammering away at our
work, we sit around and talk of the latest developments.
These things make such an impression that I can quite
understand old veterans boring everybody to death
with reminiscences. I see some forty years from now
that people will be saying: "I don't want to let old
man Gibson get hold of me and tell me all about the
war of 1914!"

This morning I received a telegram from Richard Harding Davis, who wants to join the Belgian forces. We are trying to arrange it this morning, and I expect to see him any day now.

We are going to have a lot of newspaper men in our midst. I met two more of them last night. None of them who have so far appeared speak any language but English, but they are all quite confident that they can get all the news. I look next for Palmer and Jimmy Hare and the rest of the crowd.

Maxwell, the *Telegraph* correspondent, yesterday showed me a photograph of a French bulldog that has been doing good service at Liège. His master, who is an officer in one of the forts, fastens messages in his collar and shoves him out onto the glacis. The puppy makes a blue streak for home and, as he is always sent at night, has managed so far to avoid the Germans. His mistress brings him back to the edge of town and starts him back for the fort.

The Belgian troops have so far had to dam the flood of Germans with little or no help from the allies. The Kaiser expected, so far as we can make out, to sweep through Belgium with little opposition and be fighting in France in three days! The Belgians have knocked his schedule out by twelve days already, and there is no telling how much longer they may hold out. "My military advisers" tell me that in view of the great necessity for a quick campaign in France, so as to get the army back in time to head off the Russian flood when it begins to pour over the northern frontier, the loss of this much time is equivalent to the loss of the first great battle. The moral effect is also tremendous.

The Minister to-day had a card from Omer which began: *"J'ai l'honneur de faire savoir a Votre Excellence que je suis encore toujours vivant!"* *Encore toujours* sounds as though he were pretty emphatically alive. We were all relieved to hear from him.

Villalobar, the Spanish Minister, came in after dinner—just to visit. His household is greatly upset. His cook and three footmen have gone to the war. He apologised for not inviting us to dine during these depressing days, but said he could not, as his cook was a Lucretia di Borgia. He is confident that the war is going to knock Brussels life into a cocked hat this winter. So many of the families will be in mourning, and so much poverty will come as a result of the war. Life goes on so normally now, save for the little annoyances of living under martial law, that it is hard to realise that such great changes are imminent.

Brussels, August 16, 1914.—This morning I walked out of my office and bumped into Frederick Palmer. I had no idea he was so near. Two weeks ago he was in Vera Cruz, but made a bee-line for Brussels at the first news of impending war. In the breathing spaces during the morning I got in a little visiting with him. He stayed to lunch at the Legation and so did I. In the afternoon I took him to the Foreign Office and the War Office and the Gendarmerie, and got him outfitted with passes, so that he can make a try to get towards the front. As a measure of precaution I added another *laisser-passer* to my collection, with a beautiful photograph on it. The collection grows every day.

I went to the Palace to dine with Palmer and Blount.

We had hardly got seated when in walked Richard Harding Davis and Gerald Morgan, and joined us. I had not expected Davis here so soon, but here he is. He was immaculate in dinner jacket and white linen, for war does not interfere with his dressing.

While we were dining, a lot of motors came by filled with British officers. There was a big crowd in the square, and they went crazy with enthusiasm, cheering until the windows rattled.

Brussels, August 18, 1914.—At ten in the morning I started with Frederick Palmer and Blount in the latter's car, to see whether we could get a little way out of town and get a glimpse of what was going on. We were provided with *laisser-passers* and passports and all sorts of credentials, but as a strict prohibition against sightseers has been enforced for some days, we rather doubted whether we should be able to get farther than the edge of town. Before we got back we had gone more than a hundred kilometers through the heart of things and saw a great deal more than anybody should be allowed to see. We got back to town about eight o'clock, thoroughly tired and with eyes filled with dust and cinders.

Part way out the avenue we were hailed by a soldier, who asked us for a lift as far as Tervueren. He climbed into the car beside me and rode out. The Forêt de Soignes was mournful. Quatre Bras, where the cafés are usually filled with a good-sized crowd of bourgeois, was deserted and empty. The shutters were up and the proprietors evidently gone. The Minister's house, near by, was closed. The gate was locked and the gardener's dog was the only living thing in sight. We passed our Golf Club a little farther on toward Tervueren. The old château is closed, the garden is growing rank, and the rose-bushes that were kept so scrupulously plucked and trim, were heavy with dead roses. The grass was high on the lawns; weeds were springing up on the fine tennis courts. The gardeners and other servants have all been called to the colours. Most of the members are also at the front, shoulder to shoulder with the servants. A few caddies were sitting mournfully on the grass and greeted us solemnly and without enthusiasm. These deserted places are in some ways more dreadful than the real horrors at the front. At least there is life and activity at the front.

Before we got out of town the guards began stopping us, and we were held up every few minutes until we got back to town at night. Sometimes the posts were a kilometer or even two kilometers apart. Sometimes we were held up every fifty yards. Sometimes the posts were regulars, sometimes Gardes Civiques; often hastily assembled civilians, mostly too old or too young for more active service. They had no uniforms, but only rifles, caps, and brassards to distinguish them as men in authority. In some places the men formed a solid rank across the road. In others they sat by the roadside and came out only when we hove in sight. Our *laisser-passers* were carefully examined each time we were stopped, even by many of the guards who did not understand a word of French, and strangely enough, our papers were made out in only the one language. They could, at least, understand our photographs and took the rest for granted.

When we got to the first outpost at Tervueren, the guard waved our papers aside and demanded the password. Then our soldier passenger leaned across in front of Blount and whispered "*Belgique.*" That got us through everything until midday, when the word changed.

From Tervueren on we began to realise that there was really a war in progress. All was preparation. We passed long trains of motor trucks carrying provisions to the front. Supply depots were planted along the way. Officers dashed by in motors. Small detachments of cavalry, infantry and artillery pounded along the road toward Louvain. A little way out we passed a company of scouts on bicycles. They are doing good work, and have kept wonderfully fresh. In this part of the country everybody looked tense and anxious and hurried. Nearer the front they were more calm.

Most of the groups we passed mistook our flag for a British standard and cheered with a good will. Once in a while somebody who recognised the flag would give it a cheer on its own account, and we got a smile everywhere.

All the farm houses along the road were either already abandoned or prepared for instant flight. In some places the reaping had already begun, only to be abandoned. In others the crop stood ripe, waiting for the reapers that may never come. The sight of these poor peasants fleeing like hunted beasts and their empty houses or their rotting crops were the worst part of the day. It is a shame that those responsible for all this misery cannot be made to pay the penalty— and they never can, no matter what is done to them.

Louvain is the headquarters of the King and his État-Major. The King is Commander-in-Chief of the Allied Forces operating in Belgium, and is apparently proving to be very much of a soldier. The town is completely occupied and troops line the streets, stopping all motors and inspecting papers, then telling you which way you can go. We were the only civilians on the road all day, except the Red Cross people. The big square was completely barred off from general traffic and was surrounded with grenadiers. We got through the town and stopped at the only café we could find open, where we had a bottle of mineral water and talked over what we should do next.

In Louvain there is an American theological seminary. We had had some correspondence with Monseigneur de Becker, its Rector, as to what he should do to protect the institution. At our suggestion he had established a Red Cross Hospital and had hoisted a big American flag, but still he was not altogether easy in his mind. I called on him and did my level best to reassure him, on the ground that the Germans were certainly not making war on seminaries or priests, and that if the Germans reached Louvain, all he had to do was to stay peacefully at home and wait for quiet to be restored. Most of his students were gone and some of the faculty had followed them, so his chief concern was for the library and other treasures. My arguments did not seem to have very much weight, but I left with a promise to look in again at the first opportunity and to respond to any call the Rector might make.

From the seminary we drove out the Tirlemont road, to see if we could get to that little town and see some of the fighting that was known to be going on. At the edge of the town we came to a barricade of carts, road-rollers and cobble stones, where we were courteously but firmly turned back. Everybody was anxious to make it as nice as possible for us, and one of the bright boys was brought forward to tell us in English, so as to be more convincing. He smiled deprecatingly, and said: "Verreh bad. Verreh sorreh. Oui mus' mak our office, not?" So we turned and went back to town. They had told us that *nobody* could go beyond the barricade without an order from the *Commandant de Place* at Louvain. On the way back we decided that we could at least try, so we hunted through the town until we found the headquarters of the Commandant. A fierce-looking sergeant was sitting at a table near the door, hearing requests for visés on *laisser-passers*. Everybody was begging for a visé on one pretext or another, and most of them were being turned down. I decided to try a play of confidence, so took our three cards and walked up to his table, as though there could be no possible doubt of his doing what I wanted. I threw our three *laisser-passers* down in front of him, and said in a business-like tone: "*Trois visés pour Tirlemont, S. V. P.*" My man looked up in mild surprise, viséed the three papers without a word and handed them back in less time than it takes to tell it. We sailed back to the barricade in high feather, astonished the guard with our visé, and plowed along the road, weaving in and out among ammunition wagons, artillery caissons, infantry, cavalry, bicyclists—all in a dense cloud of dust. Troops were everywhere in small numbers. Machine guns, covered with shrubbery, were thick on the road and in the woods. There was a decidedly hectic movement toward the front, and it was being carried out at high speed without confusion or disorder. It was a sight to remember. All along the road we were cheered both as Americans and in the belief that we were British. Whenever we were stopped at a barricade to have our papers examined, the soldiers crowded around the car and asked for news from other parts of the field, and everybody was wild for newspapers. Unfortunately we had only a couple that had been left in the car by accident in the morning. If we had only thought a little, we could have taken out a cartful of papers and given pleasure to hundreds.

The barricades were more numerous as we drew nearer the town. About two miles out we were stopped dead. Fighting was going on just ahead, between us and the town, and the order had been given out that *nobody* should pass. That applied to military and civilians alike, so we could not complain, and came back to Louvain, rejoicing that we had been able to get so far.

We hunted up our little café and ate our sandwiches at a table on the sidewalk, letting the house profit to the extent of three glasses of beer. We were hardly seated when a hush fell on the people sitting near. The proprietor was summoned and a whispered conversation ensued between him and a bewhiskered old man three tables away. Then Mr. Proprietor sauntered over our way with the exaggerated carelessness of a stage detective. He stood near us for a minute or two, apparently very much interested in nothing at all. Then he went back, reported to "Whiskers" and

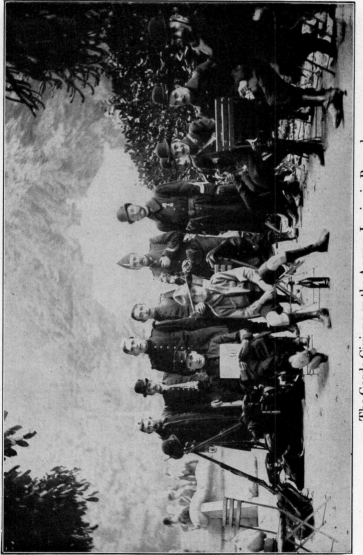

The Garde Civique on the Avenue Louise in Brussels

Types of Belgian cavalrymen

the buzz of conversation began again as though nothing had happened. After a bit the proprietor came over again, welcomed us to the city, asked us a lot of questions about ourselves, and finally confided to us that we had been pointed out as Germans and that he had listened to us carefully and discovered that we were nothing of the sort. *"J'ai très bonne oreille pour les langues,"* he said. Of course we were greatly surprised to learn that we had been under observation. Think of German spies within 200 yards of the headquarters of the General Staff! (And yet they have caught them that near.) Every active citizen now considers himself a policeman on special duty to catch spies, and lots of people suffer from it. I was just as glad the proprietor had not denounced us as spies, as the populace has a quite understandable distaste for them. I was glad the bright café proprietor could distinguish our lingo from German.

After lunch we went down to the headquarters of the General Staff, to see if we needed any more visés. We did not, but we got a sight of the headquarters with officers in all sorts of uniforms coming and going. The square was full of staff autos. The beautiful carved Hôtel de Ville is the headquarters. As we walked by, a British Major-General came down the steps, returned everybody's salutes and rolled away—a fine gaunt old type with white hair and moustache—the sort you read about in story books.

After lunch we found that there was no use in trying to get to Tirlemont, so gave that up, and inquired about the road to Diest. Everybody who was in any sort of position to know told us we could not get more than a few kilometers along the road, and that as Uhlans were prowling in that neighbourhood, we might be potted at from the woods or even carried off. On the strength of that we decided to try that road, feeling fairly confident that the worst that could happen to us would be to be turned back.

As we drew out along the road, the traffic got steadily heavier. Motors of all sorts—beautifully finished limousines filled with boxes of ammunition or sacks of food, carriages piled high with raw meat and cases of biscuit. Even dog-carts in large numbers, with the good Belgian dogs straining away at the traces with a good will, and barking with excitement. They seemed to have the fever and enthusiasm of the men and every one was pulling with all his strength. In some places we saw men pushing heavily- laden wheelbarrows, with one or two dogs pulling in front.

From Louvain on most of the barricades were mined. We could see clearly as we passed where the mines were planted. The battery jars were under the shelter of the barricade and the wire disappeared into some neighbouring wood or field. Earthworks were planted in the fields all along the lines, good, effective, well-concealed intrenchments that would give lots of trouble to an attacking force. There was one place where an important intrenchment was placed in a field of hay. The breastworks were carefully covered with hay and the men had it tied around their hats in such a way as to conceal them almost completely. This war is evidently going to be fought with some attention to detail, and with resourcefulness.

Diest itself we reached at about half past three, after having been nearly turned back six or seven times.

We were the only civilians that had turned up all day, and although our papers seemed to be all right and we could give a good account of ourselves, our mere presence was considered so remarkable that a good many of the outposts were inclined to turn us back. By virtue of our good arguments and our equally good looks, however, we did manage to get through to the town itself.

Diest is an old town which figures a good deal in the combats of the middle ages. It has a fine old church, quite large, a good Hôtel de Ville, and clean, Dutch-looking streets, with canals here and there. The whole town is surrounded with high earthworks, which constituted the fortifications, which were part of the line of forts erected by the allies after Waterloo, as a line of defence against French aggression. These forts were so numerous that Belgium in her younger days had not sufficient men to garrison them. A number of them were abandoned, finally leaving Antwerp, Liège and Namur to bear the burden. Brialmont, who built the great ring forts at Liège, wanted to build modern fortifications at Diest, but could not get those holding the purse-strings to see things his way.

Diest was attacked by Germans about three days ago. They wanted to take the old fortifications so as to control the road and use the place as a base of operations. It could hardly be called a big battle, but was more probably in the nature of a reconnaissance in force with four or five regiments of cavalry. This part of Belgium is the only place on the whole field of operations where cavalry can be used and they are certainly using it with a liberal hand, probably in attempt to feel out the country and locate the main body of opposing troops. They have got into a lot of trouble so far, and I am sure they have not yet located the main bodies of the allied armies.

The shops were all closed and most of the people were sitting on the sidewalk waiting for something to turn up. Some of them had evidently been to America, and we had an ovation all the way in. The Grande Place was filled with motors and motor trucks, this evidently being a supply depôt. We had some of the local mineral water and talked with the people who gathered round for a look at the *Angliches.*

They were all ready for anything that might come, particularly Prussians. In the old days the Uhlans spread terror wherever they appeared, to burn and shoot and plunder. Now they seem to arouse only rage and a determination to fight to the last breath. There was a little popping to the north and a general scurry to find out what was up. We jumped in the car and made good time through the crowded, crooked little streets to the fortifications. We were too late, however, to see the real row. Some Uhlans had strayed right up to the edge of town and had been surprised by a few men on the earthworks. There were no fatalities, but two wounded Germans were brought into town in a motor. They were picked up without loss of time and transported to the nearest Red Cross hospital.

Cursing our luck we started off to Haelen for a look at the battlefields. Prussian cavalry made an attack there the same day they attacked Diest, and their losses were pretty bad.

At one of the barricades we found people with Prussian lances, caps, haversacks, etc., which they were

perfectly willing to sell.   Palmer was equally keen to
buy, and he looked over the junk offered, while some
two hundred soldiers gathered around to help and
criticise.   I urged Palmer to refrain, in the hope of
finding some things ourselves on the battlefield.   He
scoffed at the idea, however.   He is, of course, an old
veteran among the war correspondents, and knew
what he was about.   He said he had let slip any number of opportunities to get good things, in the hope of
finding something himself, but there was nothing doing
when he got to the field.   We bowed to his superior
knowledge and experience, and let him hand over an
English sovereign for a long Prussian lance.   I decided
to do my buying on the way home if I could find
nothing myself.

The forward movement of troops seemed to be headed
toward Diest, for our road was much more free from
traffic.   We got into Haelen in short order and spent
a most interesting half hour, talking to the officer in
command of the village.   As we came through the
village we saw the effect of rifle fire and the work of
machine guns on the walls of the houses.   Some of them
had been hit in the upper story with shrapnel and were
pretty badly battered up.   The village must have been
quite unpleasant as a place of residence while the row
was on.   The commanding officer, a major, seemed
glad to find some one to talk to, and we stretched our
legs for half an hour or so in front of his headquarters
and let him tell us all about what had happened.   He
was tense with rage against the Germans, whom he
accused of all sorts of barbarous practices, and whom
he announced the allies must sweep from the earth.

He told us that only a few hours before a couple of Uhlans had appeared in a field a few hundred yards from where we were standing, had fired on two peasant women working there, and then galloped off. Everywhere we went we heard stories of peaceful peasants being fired on. It seems hard to believe, but the stories are terribly persistent. There may be some sniping by the non-combatant population, but the authorities are doing everything they can to prevent it, by requiring them to give up their arms and pointing out the danger of reprisals.

Before we moved on, our officer presented me with a Prussian lance he had picked up on the battlefield near Haelen. We got careful directions from him for finding the battlefield and set off for Loxbergen, where the fight had taken place the day before. The run was about four kilometers through little farms, where the houses had been set on fire by shrapnel and were still burning. The poor peasants were wandering around in the ruins, trying to save odds and ends from the wreck, but there was practically nothing left. Of course they had had to flee for their lives when the houses were shelled, and pretty much everything was burned before they could safely venture back to their homes.

We had no difficulty in locating the field of battle when we reached it. The ground was strewn with lances and arms of all sorts, haversacks, saddle bags, trumpets, helmets and other things that had been left on the ground after the battle. There were a few villagers prowling around, picking things up, but there were enough for everybody, so we got out and gathered about fifteen Prussian lances, some helmets and other odds and ends that would serve as souvenirs for our friends in Brussels. As everybody took us for English, they were inclined to be very friendly, and we were given several choice trophies to bring back. While we were on the field, a German aeroplane came soaring down close to us and startled us with the sharp crackling of its motor. It took a good look at us and then went its way. A little farther along, some Belgian troops fired at the aeroplane, but evidently went wide of their mark, for it went unconcernedly homeward. We wandered through the ruins of some old farms and sized up pretty well what must have happened. The Germans had evidently come up from the south and occupied some of the farmhouses along the road. The Belgians had come down from the north and opened fire on the houses with rapid-fire guns, for the walls were riddled with small holes and chipped with rifle fire. Then shrapnel had been brought into play, to set the houses on fire and bring the German troops out into the open. Then they had charged the Belgians across an open field and apparently with disastrous results. Part of the ground was in hay which had already been harvested and piled in stacks, the rest was in sugar beets. The Prussians had charged across the field and had come upon a sunken road into which they fell helter-skelter without having time to draw rein. We could see where the horses had fallen, how they had scrambled to their feet and tried with might and main to paw their way up on the other side. The whole bank was pawed down, and the marks of hoofs were everywhere. The road was filled with lances

and saddles, etc.   All through the field were new-made
graves.   There was, of course, no time for careful
burial.   A shallow trench was dug every little way—a
trench about thirty feet long and ten feet wide.   Into
this were dumped indiscriminately Germans and Belgians and horses, and the earth hastily thrown over
them—just enough to cover them before the summer
sun got in its work.   There were evidences of haste;
in one place we saw the arm of a German sergeant
projecting from the ground.   It is said that over
three thousand men were killed in this engagement,
but from the number of graves we saw I am convinced
that this was a good deal overstated.   At any rate
it was terrible enough; and when we think that this
was a relatively unimportant engagement, we can form
some idea of what is going to happen when the big
encounter comes, as it will in the course of a few days
more.   It is clear that the Germans were driven
off with considerable losses, and that the Belgians
still hold undisputed control of the neighbourhood.
There were a few scattered Uhlans reconnoitering near
by, but they were not in sufficient numbers to dare
to attack.

After gathering our trophies we were ready to start
for home; and it was well we should, for it was getting
rather late in the afternoon and we had a long trip
ahead of us with many delays.

Soon after leaving Haelen, on our way back we met
a corps of bicycle carabiniers who were rolling along
toward Haelen at top speed.   The officer in command
held us up and asked us for news of the country we
had covered.   He seemed surprised that we had not

seen any German forces, for he said the alarm had been
sent in from Haelen and that there were strong forces
of Belgians on the way to occupy the town and be
ready for the attack. When he had left us, we ran
into one detachment after another of infantry and
lancers coming up to occupy the little village.

When we got to the barricade at the entrance to
Diest, the soldiers of the guard poured out and began
taking our trophies out of the car. We protested
vigorously, but not one of them could talk anything
but Walloon—and French was of no use. Finally,
a corporal was resurrected from somewhere and came
forth with a few words of French concealed about his
person. We used our best arguments with him, and
he finally agreed to let a soldier accompany us to the
town hall and see what would be done with us there.
The little chunky Walloon who had held us up at the
barrier climbed in with great joy, and away we sped.
The little chap was about the size and shape of an
egg with whopping boots, and armed to the teeth.
He had never been in a car before, and was as delighted
as a child. By carefully piecing words together through
their resemblance to German, we managed to have
quite a conversation; and by the time we got to the
Grande Place we were comrades in arms. I fed him
on cigars and chocolate, and he was ready to plead
our cause. As we came through the streets of the
town, people began to spot what was in the car and
cheers were raised all along the line. When we got
to the Hôtel de Ville, the troops had to come out to
keep back the curious crowd, while we went in to inquire
of the officer in command as to whether we could keep

our souvenirs. He was a Major, a very courteous and patient man, who explained that he had the strictest orders not to let anything of the sort be carried away to Brussels. We bowed gracefully to the inevitable, and placed our relics on a huge pile in front of the Hôtel de Ville. Evidently many others had met the same fate, for the pile contained enough trophies to equip a regiment. The Major and an old fighting priest came out and commiserated with us on our hard luck, but their commiseration was not strong enough to cause them to depart from their instructions.

The Major told us that they had in the Hôtel de Ville the regimental standard of the Death's Head Hussars. They are keeping it there, although it would probably be a great deal safer in Brussels. Unfortunately the room was locked, and the officer who had the key had gone, so we could not look upon it with our own eyes.

Heading out of town, a young infantryman held us up and asked for a lift. He turned out to be the son of the President of the Court of Appeals at Charleroi. He was a delicate looking chap with lots of nerve, but little strength. His heavy infantry boots looked doubly heavy on him, and he was evidently in a bad way from fatigue. He had to rejoin his regiment which was twelve miles along the road from Diest, so we were able to give him quite a boost. He asked me to get word to his father that he wanted to be given a place as chauffeur or aviator, and in any other place that would not require so much foot work. There must be a lot of this sort. We finally landed him in the bosom of his company and waved him a goodbye.

By this time it was twilight, and the precautions of the guards were redoubled. A short way out from Louvain, a little Walloon stepped out from behind a tree about a hundred yards in front of us and barred the way excitedly. We were going pretty fast and had to put on emergency brakes, and skid up to him with a great smell of sizzling rubber. He informed us that papers were no good any more; that we must know the password, or go back to Louvain for the night. This he communicated to us in his best Walloon, which we finally understood. Blount started to tell him that we did not know, as the word had been changed since we left; but in one of my rare bursts of resourcefulness I thought to try a ruse, so leaned forward very confidently and gave him the password for the morning—"*Belgique.*" With a triumphant look, he shook his head and countered: "No, *Haelen!*" He had shown the travellers from the outside world that he knew more than they did, and he was without any misgivings as to what he had done, and let us proceed without further loss of time. We got all the way back to Tervueren with this password, which was all that saved us from spending the night in Louvain and getting back nobody knows when. Nearly opposite the Golf Club we were stopped with the tidings that the word was no longer good, but that if we had satisfactory papers we could get into town. For some reason the password had evidently been changed since we left Louvain, so we got through with rare luck all along the line.

We rolled up to the Legation a few minutes before eight o'clock, and found that there was a great deal of anxiety about us.   Cheerful people had been spreading the news all day that if we fell into the hands of the Germans they would hold us as hostages, as they did the Bishop and Mayor of Liège.   They probably would if they had caught us, but they did not catch us.

Palmer was pleased at the amount we saw.   It was by rare good luck that we got through the lines and we were probably the last who will get so far. To-day all *laisser-passers* have been canceled, and nobody can set foot out of town to the east.   It gave us a pretty good idea before we got through as to how the troops must be disposed.   I came within an ace of putting off our trip for a day or two.   If I had, it would have cut me out of seeing anything.

As usual, when I go out, the lid had blown off the Legation and the place was in a turmoil.   During the afternoon the Government had decided to move to Antwerp and take refuge in the *enceinte*.   The Queen, the royal children and some of the members of the Government left at eight o'clock, and this morning more of them left.   Most of the Diplomatic Corps have gone, and will have so much time to think of their troubles that they will be more uncomfortable than we are.   The Spanish Minister will stay on and give us moral support.

Brussels, August 19, 1914.—Yesterday morning began with a visit from our old friend, Richard Harding Davis, who was still quite wroth because I had not waited for him to arrange for his passes and go with me on my trip.   If we had, there would have been no trip, as he was not equipped until afternoon. After lunch he started off boldly for Namur, but got turned back before he reached Wavre, where there had been a skirmish with Uhlans. He was sore and disgusted.

While he was in my office, another troop arrived composed of Irwin Cobb, John McCutcheon, the cartoonist, Lewis and a few others. Later in the day, Will Irwin came in with news that he was closely followed by others. McCutcheon is a great friend of the Minister, and makes this his headquarters.

The Minister took them out to get *laisser-passers*. While they were away, Sir Francis Villiers came in and showed me a telegram from the Foreign Office, stating that British newspapers and news associations had been requested to recall their correspondents, as they had already done great harm by the news they had given out. He was also to request the Belgian Government to refuse permits of any sort to the press, and get all foreign correspondents out of the country. The Belgian Government realised the importance of this, and has consequently shut down the lid tight.

There was supposed to have been a fair-sized cavalry engagement near Charleroi, in which six regiments of German cavalry were chewed up. We have no details, but it looked as though they were lured into a trap. Practically no news of the operations is leaking out. It looks as though Kitchener had remarked, "We will go into that house where William Hohenzollern is breaking the furniture, and we will close the door and pull down the blinds, and when we get through, we will come out and tell people about it."

Yesterday was just a day of work with a great deal of beating people on the back and assuring them that their lives are not in danger just because the Court has gone to Antwerp. They all seem to be convinced that their throats are going to be cut immediately.

This morning we had the usual deluge of newspaper men and correspondents. The Minister went off with the Spanish Minister to call on the military authorities, who are the only ones with whom we now have any relations, and while he was gone, Sir Francis came in and announced that he had been ordered to leave for Antwerp and place his Legation and British interests under our charge. The news is that the German cavalry in considerable force is marching toward Brussels. The military authorities are getting ready to defend the city, which is quite a futile proceeding, as the available forces are inadequate, so that the only result will be that a lot of innocent people will be killed quite incidentally. The Governor expects to resist about as far as the ring of inner boulevards, which are about four blocks farther in than we are. Our street is probably one of the principal ones by which the Germans would enter. A hundred yards farther out there is a big railroad barricade, where a stand would probably be made, so that our Legation would undoubtedly get a fair share of the wild shots from both sides. The cellar is being made ready for occupancy during the shindy, if it comes. The Burgomaster came in to say that he had a house prepared for our occupancy in the safe part of town; but we were not prepared to abandon the Legation and declined with sincere thanks for his thoughtfulness.

I went over and saw Sir Francis and the Legation staff just as they were leaving. They refused to have their plans upset by any little thing like a German advance, so had their lunch peacefully at the usual hour and then left in motors.

At seven o'clock Cobb, McCutcheon, and the rest of the crowd, were due at my house, so I gathered up the Minister, the Consul-General, and Blount, and repaired thither. Davis and Morgan turned up a little late, but nothing has been heard of the rest of the crowd so far—10:30 P. M. They were to have dined here, but have not appeared or sent word.

Crowds of people are pouring in from the east in all stages of panic, and some small forces of cavalry have also retreated into the city, looking weary and discouraged. There has evidently been a rout. Further than that, we know nothing so far.

Several of the wives of high Belgian officials have come in this evening, having received word from their husbands to put themselves under our protection. There is nothing we can do for them, particularly at this time.

Brussels, August 20, 1914.—To-day has been one full of experience and the end is not yet. Last night there was a great stir in the streets, and crowds of people and weary-looking soldiers. At the Palace Hotel I found the usual collection of diplomats and some other people whom I knew, and from the crowd I elicited the fact that there had been some sort of rout of Belgian forces near Louvain, and the soldiers were falling back. That was about all they knew. I started back to the upper town in the hope of finding some news at the Porte de Namur. On the way up the hill I was stopped by half a dozen groups of Gardes Civiques and soldiers, who asked me to take them to Ghent. They were so excited and in such a hurry that they could hardly be made to realize that the car was not liable to seizure. I took advantage of the opportunity to get a little first-hand news, and learned that they had been driven back all along the line and were ordered to retreat to Ghent by any means they could find. There were no trains available; nobody seemed to know why. The last group that I talked with said that the vanguard of the German cavalry was only about fifteen miles out of town and would be in this morning. They were all tremendously excited and did not dally by the wayside to chat about the situation with me. I can't say that I blame them, particularly in view of what I have seen since.

At the Porte de Namur I found that the Garde Civique in Brussels had been ordered to disband and that the plan for the defense of the city had been completely abandoned. It was the wise thing to do, for there was no hope of defending the town with the small force of Gardes at the disposal of the military governor. It would have been quite futile and would have entailed a big loss of innocent civilian life. The governor wanted to do it purely as a matter of honour, but he would have paid for it heavily and could not have accomplished anything beyond delaying the Germans for an hour or two. The Garde Civique was furious, however, at the idea of not being able to make a stand. There was a demonstration, but the cooler heads prevailed, and the men withdrew to their homes.

German supply train entering Brussels

German infantry entering Brussels

I was out by seven this morning and looked about for news before coming to the Legation. I found that the Germans were steadily advancing and that the vanguard was about seven kilometers out of the city. They expected to begin the triumphal march about eleven. The Garde Civique had disappeared from the streets and there were very few police to be found. The shops were closed, shutters down on all houses, and posters everywhere with the proclamation of the Burgomaster urging the people to refrain from hostile acts. It was an abandoned and discouraged-looking city. On the boulevards there were long lines of high carts bringing in the peasants from the surrounding country. They are great high-wheeled affairs, each drawn by a big Belgian draught horse. Each cart was piled high with such belongings as could be brought away in the rush. On top of the belongings were piled children and the old women, all of whom had contrived to save their umbrellas and their gleaming, jet-black bonnets, piled with finery. Those who could not find places in the carts walked alongside, some of them carrying other belongings that could not be put on the carts. It was the most depressing sight so far. Lots of them were crying; all looked sad and crushed. Every one of them was probably without enough money for a week's living. Even those who have money in the banks cannot get it out at this time. They have no place to go to here and have a bad prospect even if this part of the campaign is finished quickly and they are soon able to return to their homes. Their crops are rotting in the ground and many of their homes are already in ruins. That is the hard side of the war—lots harder than the men who go out and have at least a fighting chance for their lives.

When I got down to the Legation I found that the telegraph and telephone communication had been cut off. The train service is abandoned and we are completely isolated from the outside world. I did not think it would come so soon and only hope that before we were cut off the news was allowed to get out that there would be no fighting in the city.

I had a lot of errands to do during the morning and kept both motors busy. I found time to get up signs on my door and that of M. de Leval, warning all comers that both places were inviolate. That was in anticipation of quartering of troops on private citizens, which has not been done.

We got word that the Spanish Minister had some news, so I went over to see him. He had heard from the Burgomaster as to the plans for the entry of the troops, and wanted to pass it along to us. The commanding general, von Jarotzky, was already at the edge of the city, on the Boulevard Militaire, and was expecting to start into town at one o'clock. He was to march down the Chaussée de Louvain, the boulevards, and out the other side of the city, where his men were to be encamped for the present. Other forces, comparatively small, were to occupy the railway stations and the Grande Place. At the Hôtel de Ville he was to establish the headquarters of the staff and administer the city government through the regularly constituted authorities. It was all worked out to a nicety, even to the exact measures for policing the line of march.

As the Garde Civique was withdrawn, the prisoners in the German Legation knew that there was something in the air and ventured forth into the light of day. They were not long in learning just what had taken place, and called upon us to express their thanks for what we had done for them. I suppose they will be trotting away for their own country before there is a chance to lock them up again. It must be pretty dismal for them to be locked up without any news of the outside world when they don't know whether their armies are victorious or badly beaten.

As I was about to start to see the triumphal entry, the Spanish Minister came along with his flag flying from his motor, and bade us to go with him. We made off down the Boulevard and drew up at the Italian Legation—two motors full of us; the whole staff of the Spanish Legation and ourselves. The Italian Minister bade us in to watch the show, which we had intended he should do. This did not work out well, so M. de Leval and I started off down the street together. The first of the Germans appeared as we stepped out the front door, and we saw that they were not coming over the route that had been originally planned. Instead, they were heading down the hill into the lower town. They proved to be the troops that were to occupy the Grande Place and guard the headquarters of the staff at the Hôtel de Ville. We cut across through side streets and came upon them as they were passing Ste. Gudule. There was a sullen and depressed crowd lining the streets, and not a sound was to be heard. It would have been better had the crowd been kept off the streets, but they behaved wonderfully well.

A large part of the reason for bringing the German troops through here was evidently to impress the populace with their force and discipline. It was a wonderful sight, and one which I never expect to see equaled as long as I live. They poured down the hill in a steady stream without a pause or a break; not an order was shouted nor a word exchanged among the officers or men. All the orders and signals were given by whistles and signs. The silence was a large element of the impressiveness.

These troops had evidently been kept fresh for this march, and I should not be at all surprised if it should prove that they had not seen any fighting. If they have suffered any losses, they have closed up their ranks with wonderful precision and show none of the signs of demoralisation. They had clearly been at great pains to brush up and give the appearance of freshness and strength. Nearly all the men were freshly shaven, and their uniforms had been brushed and made as natty and presentable as possible. They swaggered along with a palpable effort to show that they were entirely at home, and that they owned the place. The officers looked over the heads of the crowd in their best supercilious manner, and the men did their best to imitate their superiors.

First came some lancers—a couple of battalions, I should think; then there was a lot of artillery, rapid-fire guns and field pieces. Then more cavalry and a full regiment of infantry. When the last contingent of cavalry came along, they burst into song and kept it up steadily. There was a decidedly triumphant note, and the men looked meaningly at the crowd, as

much as to say: "Now do you realise what your little
army went up against when it tried to block us?" It
seemed to me pretty rough to rub it in on them by
singing songs of triumph as they rode into an undefended city. If they had been attacked and had
succeeded in driving the invader back into his own
capital, it would be understandable; but it seemed to
me rather unnecessary to humiliate these people after
trampling on their poor country and slaughtering half
their army. It was more than de Leval could stand,
so I walked home with him to the Legation.

When we got back to the Legation I decided that
I ought to see all I could, so Blount and I went back
in his car. First we worked our way through to the
lower town and got a look at the Grande Place.
There were a little more than two full battalions resting
there, with their field pieces parked at the lower end
of the square. Small squads were being walked around
doing the goose step for the delectation of the *bons
Bruxellois*, who were kept a block away up the side
streets leading to the square. The men had their
arms stacked in the centre of the square, and were
resting hard—all but those who were supplying the
spectacle.

From there we went down to Luna Park, an amusement place on the edge of the city. The stream was
pouring by there just as steadily as it had earlier in
the afternoon. We watched the passing of great
quantities of artillery, cavalry and infantry, hussars,
lancers, cyclists, ambulance attendants, forage men,
and goodness only knows what else.

I have never seen so much system and such equipment. The machine is certainly wonderful; and, no matter what is the final issue of the war, nobody can deny that so far as that part of the preparation went, the Germans were hard to beat. The most insignificant details were worked out, and all eventualities met with promptness. The horses were shod for a campaign in the country, and naturally there was a lot of slipping on the smooth cobble pavements. The instant a horse went down there was a man ready with a coarse cloth to put under his head, and another to go under his forefeet, so that he would have some grip when he tried to get up and would not hurt himself slipping and pawing at the cobbles. The moment he fell, all hands rushed to the rescue so effectively that he was on his feet again in no time, and the procession was barely arrested. The men's kits were wonderfully complete and contained all sorts of things that I had never seen or heard of, so I turned for explanation to Davis, who had come along and was lost in admiration of the equipment and discipline. He said he had been through pretty much every campaign for the last twenty years, and thought he knew the last word in all sorts of equipment, but that this had him staggered. I began asking him what a lot of things were for, and he frankly admitted that he was as much in the dark as I was.

A great many of the officers wore, upon their chests, great electric searchlights attached to batteries in their saddle-bags. These are useful when on the march at night, and serve to read sign-posts and study maps, etc.

The supply trains were right with the main body of the troops, and were also carefully equipped for purposes of display. The kitchens were on wheels, and each was drawn by four horses. The stoves were lighted and smoke was pouring from the chimneys. The horses were in fine shape and in huge numbers.

The troops marched down the right side of the boulevard, leaving the left side free. Up and down this side dashed officers on horseback, messengers on motor-cycles and staff officers in military cars. There were no halts and practically no slacking of the pace, as the great army rolled in.

Here and there came large motor trucks fitted out as cobblers' shops, each with a dozen cobblers pounding industriously away at boots that were passed up to them by the marching soldiers. While waiting for repairs to be made, these soldiers rode on the running board of the motor, which was broad enough to carry them and their kits.

After watching them for a while, we moved back to the Boulevard, where we found the Minister with the ladies of the family who had been brought out to watch the passing show. We had hesitated to bring them out at the beginning for fear that there might be riots, or even worse, precipitated by the foolhardy action of some individual. Fortunately, there was nothing of the sort, and while the reception given the troops was deadly sullen, they were offered no affronts that we could see. The entry was effected quietly, and perfect order has prevailed ever since.

Afterwards we drove out to the country and watched the steady stream nearer its source; still pouring in, company after company, regiment after regiment, with apparently no end in sight. We watched until after seven, and decided that the rest would have to get in without our assistance. On the way back a German monoplane flew over the city, and, turning near the Hôtel de Ville, dropped something that spit fire and sparks. Everybody in the neighbourhood let out a yell and rushed for cover in the firm belief that it was another bomb such as was dropped in Namur. It dropped, spitting fire until fairly near the spire of the Hôtel de Ville, when it burst into ten or a dozen lights like a Roman candle—evidently a signal to the troops still outside the city—perhaps to tell them that the occupation had been peacefully accomplished. We learned afterward that the Minister and Villalobar were riding down the hill and the infernal machine seemed right over their car, giving them a nice start for a moment. When I got back to the Legation, I found that the Minister had gone with Villalobar to call on the Burgomaster and the German General. They found the old gentleman in command at the city hall, carrying on the government through the Bourgomaster, who has settled down with resignation to his task. He is tremendously down in the mouth at having to give up his beautiful Grande Place to a foreign conqueror, but he has the good sense to see that he can do more good for his country by staying there and trying to maintain order than by getting out with a *beau geste*.

The first thing the General did was to excuse himself and go to take a bath and get a shave, whereupon he reappeared and announced his readiness to proceed to the discussion of business.

The General said that he had no intention of occupying the town permanently or of quartering soldiers, or otherwise bothering the inhabitants. He was sent there to keep open a way so that troops could be poured through toward the French frontier. They expect to be several days marching troops through, and during that time they will remain in nominal control of the city. Judging from this, there must be a huge army of them coming. We shall perhaps see some of them after the big engagement, which is bound to take place soon, as they get a little nearer the French frontier.

Brussels has not been occupied by a foreign army since Napoleon's time, and that was before it was the capital of a free country. It has been forty-four years since the capital of a European Power has had hostile troops marching in triumph through its streets, and the humiliation has been terrible. The Belgians have always had a tremendous city patriotism and have taken more pride in their municipal achievements than any people on earth, and it must hurt them more than it could possibly hurt any other people. The Burgomaster, when he went out to meet General von Jarotzky, declined to take his hand. He courteously explained that there was no personal affront intended, but that under the circumstances he could hardly bring himself to offer even such a purely perfunctory manifestation of friendship. The old General, who must be a good deal of a man, replied quietly that he entirely understood, and that under similar circumstances he would probably do the same. The two men are on exceedingly workable terms, but I don't believe they will exchange photographs after the war is over. Poor Max was going to spend the night at the Hôtel de

Ville. Most of his assistants cleared out for the night, but he could not bring himself to leave the beautiful old building entirely in control of the enemy. He curled up and slept on the couch in his office, just for the feeling it gave him that he was maintaining some sort of hold on the old place.

The Minister arranged to have his telegrams accepted and transmitted without loss of time, so we shall soon get word home that we are still in the land of the living. We wrote out our message and sent it off right after dinner, but Gustave brought it back, saying that the telegraph office was closed and that he could find no one to whom he could hand his bundle of messages. Evidently the orders for the re-opening of the place did not get around in time for our purposes. We shall try again the first thing in the morning, and hope that some of the newspaper men will have succeeded in getting their stuff out in some other way. They were around in force just after dinner and wild to get an O.K. on their stuff, so that it could be sent. The General had said that he wanted the Minister's O.K. on the men themselves, and that he himself would approve their messages after having them carefully read to him. He gave them an interview on alleged German atrocities and will probably let them send through their stories if they play that up properly.

After dinner I started out on my usual expedition in search of news. I found the Foreign Office closed, and learned upon inquiry that the few remaining men who had not gone to Antwerp were at home and would not be around again for the present—thus we have no dealings through the Foreign Office, but must do the best we can with the military authorities. I went down to the Palace Hotel on the chance of picking up a little news, but did not have much luck. The restaurant was half filled with German officers, who were dining with great gusto. The Belgians in the café were gathered just as far away as possible, and it was noticeable that instead of the usual row of conversation, there was a heavy silence brooding over the whole place.

August 21, 1914.—So far as we can learn we are still as completely cut off from the outside world as we were yesterday. The General promised the Minister that there would be no difficulty in sending his telegrams, either clear or in cipher, but when we came to sending them off, it was quite another story.

The first thing this morning I made an attempt to hand them in, but found all the telegraph offices closed. At ten o'clock I went down to the Hôtel de Ville to see the General, who has taken over the duties of Military Governor, and see what was the matter. He was away somewhere and so was the Burgomaster, so I contented myself with seeing one of the Echevins, whom I had met a number of times. He could not do anything about it on his own responsibility, but made a careful memorandum and said that he would take it up with the General, through the Mayor, when they both got back. I also asked for *laisser-passers* for everybody in the shop, and he promised to attend to that.

By lunch time we had received no answer from General von Jarotzky, so I got in the motor with my pocket full of telegrams and went down to the Hôtel de Ville once more. It is a depressing sight. The Grande Place, which is usually filled with flower venders and a mass of people coming and going, is almost empty. At the lower end there are parked a number of small guns; in the centre, some camp kitchens, with smoke rising from the chimneys. The courtyard of the Hôtel de Ville itself, where so many sovereigns have been received in state, was filled with saddle-horses and snorting motors. The discarded uniforms of the Garde Civique were piled high along one side, as if for a rummage sale. Beer bottles were everywhere. In the beautiful Gothic room, hung with the battle flags of several centuries, there are a hundred beds—a dormitory for the officers who are not quartered at the neighbouring hotels.

The marvelous order and system which so compelled our admiration yesterday were not in evidence. There were a lot of sentries at the door and they took care to jab a bayonet into you and tell you that you could not enter; but any sort of reply seemed to satisfy them, and you were allowed to go right up to the landing, where the General had established himself in state at a couple of huge tables. Here confusion reigned supreme. There were staff officers in abundance, but none of them seemed to have the slightest authority, and the old man had them all so completely cowed that they did not dare express an opinion or ask for a decision. The General himself is a little, tubby man, who looks as though he might be about fifty-five; his face is red as fire when it is not purple, and the way he rages about is enough to make Olympus tremble.

The crowd of frightened people who came to the Hôtel de Ville for *laisser-passers* and other papers, all found their way straight to his office; no one was on hand to sort them out and distribute them among the various bureaus of the civil administration. Even the staff officers did very little to spare their chief and head off the crowd. They would come right up to him at his table and shove a *pièce d'identitè* under his nose, with a tremulous request for a visé; he would turn upon them and growl, *"Bas bossible; keine Zeit; laissez mois dranquille, nom de D——!"* He switched languages with wonderful facility, and his cuss words were equally effective in any language that he tried. Just as with us, everyone wanted something quite out of the question and then insisted on arguing about the answer that they got. A man would come up to the General and say that he wanted to get a pass to go to Namur. The General would say impatiently that it was quite impossible, that German troops were operating over all that territory and that no one could be allowed to pass for several days. Then Mr. Man would say that that was no doubt true, but that *he* must go because he had a wife or a family or a business or something else that he wanted to get to. As he talked, the General would be getting redder and redder, and when about to explode, he would spring to his feet and advance upon his tormentor, waving his arms and roaring at him to get the —— out of there. Not satisfied with that, he invariably availed of the opportunity of being on his feet to chase all the assembled crowd down the stairs and to scream at all the officers in attendance for having allowed all this crowd to gather.

Then he would sit down and go through the same performance from the beginning. I was there off and on for more than two hours, and I know that in that time he did not do four minutes' continuous, uninterrupted work. Had it not been for the poor frightened people and the general seriousness of the situation, it would have been screamingly funny and worth staying indefinitely to see.

I had my share of the troubles. I explained my errand to an aide-de-camp and asked him to see that proper instructions were given for the sending of the telegrams. He took them and went away. Then after a few minutes he came gravely back, clicked his heels, and announced that there was no telegraph communication with the outside world and that he did not know when it would be reëstablished. I asked him to go back to the General, who in the meantime had retreated to the Gothic room and had locked himself in with a group of officers. My friend came back again, rather red in the face, and said that he had authority to stamp my telegrams and let them go. He put the rubber stamp on them and said I could take them. I said that was all very well, but where could I take them, since the telegraph offices were closed. He went off again and came back with the word that the office in the central bureau was working for official messages. I got into the motor with the Italian Secretary, who had a similar task, and together we went to the central bureau. It was nailed up tight, and the German sentries on guard at the door swore to us by their *Ehrenwort* that there was absolutely nothing doing.

Back we went to the Hôtel de Ville. Our friend, the aide-de-camp, had disappeared, but we got hold of another and asked him to inform himself. He went away and we spent a few minutes watching the General blow up everybody in sight; when the aide-de-camp came back, he smilingly announced that there was no way of getting the messages out on the wire; that the best thing we could do would be to send a courier to Holland and telegraph from there. I told him to go back and get another answer. When he came back next time, he had the glad news that the office had really been established in the post office and that orders had been sent over there to have our cables received and sent at once. Away we went again, only to find that the latest bulletin was just as good as the others; the post office was closed up just as tight as the other office, and the sentries turned us away with a weary explanation that there was not a living soul inside, as though they had explained it a thousand times since they had been on duty.

By this time the wild-goose chasing was getting a little bit monotonous, and when we got back to the headquarters, I announced with some emphasis to the first aide-de-camp that I could reach, that I did not care to do any more of it; that I wanted him to get me the right information, and do it right away, so that I should not have to go back to my chief and report any more futile errands. He went away in some trepidation and was gone some time. Presently the General came out himself, seething in his best manner.

"*A qui tout ce tas de depeches?*" roars he.

"*A moi,*" says I.

He then announced in a voice of thunder that they were all wrong and that he was having them rewritten. Before I could summon enough breath to shout him down and protest, he had gone into another room and slammed the door. I rushed back to my trusty aide-de-camp and told him to get me those telegrams right away; he came back with word that they would be sent after correction. I said that under no circumstances could they send out a word over the signature of the American Minister without his having written it himself. He came back and said that he could not get the cables. I started to walk into the office myself to get them, only to bump into the General coming out with the messages in his hand. He threw them down on a table and began telling a young officer what corrections to make on the telegraph form itself. I protested vigorously against any such proceeding, telling him that we should be glad to have his views as to any errors in our message, but that he could not touch a letter in any official message. At this stage of the game he was summoned to the office of the Burgomaster and rushed off with a string of oaths that would have made an Arizona cow-puncher take off his hat. The young officer started calmly interlining the message, so I reached over and took it away from him, with the statement that I would report to my chief what had happened. He was all aflutter, and asked that I remain, as the General would not be long. I could not see any use in waiting longer, however, and made as dignified a retreat as possible under the circumstances. There were a number of cables in the handful I had carried around that were being sent in the interest

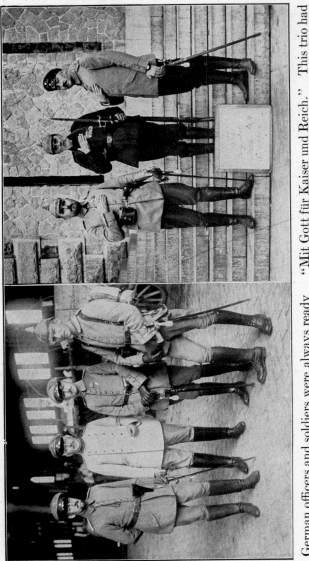

German officers and soldiers were always ready to oblige by posing for the camera. "Mit Gott für Kaiser und Reich." This trio had a mania for being photographed

Count Guy d'Oultremont, Adjutant of the Belgian Court.   From left to right: Colonel DuCane, Captain Ferguson, and Colonel Fairholme   French howitzer in the background

of the German Government and of German subjects, and I took good care to tell the young man that while we were glad to do anything reasonable for them or for their people, we had stood for a good deal more than they had a right to expect, and that these cables would stay on my desk until such time as they got ready to make a proper arrangement for our communications. Now we shall settle down and see what happens next.

When I got back to the Legation I found the Argentine and Brazilian Ministers and the Mexican Chargé d'Affaires waiting to hear the news of my mission. I was rather hot under the collar, and gave an unexpurgated account of what had happened. By this time I was beginning to see some of the humor in the situation, but they saw nothing but cause for rage, and left in a fine temper.

Just to see what would happen, we then proceeded to put our cable in its original form into cipher, and send it back to the General with a written request that it be sent immediately to Washington. It will be interesting to see what reply he makes. The Spanish Minister left some telegrams with him last night to be sent, and is quite sure that they were held up, as he has received no answers to any of them. To-morrow he expects to put on his uniform and make a solemn official call on von Jarotzky to demand that he be granted free communication with his government.

During the afternoon a lot of correspondents came in and gave an amusing account of what the General had done for them. He had received them cordially and had given them a very pleasing interview, making an extended statement about the alleged German atrocities. Could they send their messages through to their papers? Certainly! Of course the General would have to read the stories and approve the subject matter. Naturally! The boys sat down in great enthusiasm and wrote out their stories, giving full credit to the German army for the orderly way they got in, the excellence of their appearance and behaviour, and the calm that prevailed in the city. They took these messages back and let the old chap read them. He plowed his way carefully through them and expressed his great satisfaction at the friendly expressions of approval. He put his O.K. on them and handed them back with the remark that they might send them. The boys ventured to inquire how. "Oh," said the General, "you can either send a courier with them to Holland or to Germany and have them telegraphed from there." Whereupon he rose and, bowing graciously, left the bunch so flabbergasted that they did not wake up until he was gone. He was most amiable and smiling and got away with it.

The General commanding the forces now coming through—von Arnim—got out a proclamation to-day which was posted in the streets, warning the inhabitants that they would be called upon for supplies and might have troops quartered upon them, and that if they ventured upon hostile acts they would suffer severely.

PROCLAMATION.

BRUSSELS, August 20, 1914.

German troops will pass through Brussels to-day and the following days, and will be obliged by circumstances to call upon the city for lodging, food, and supplies. All these requirements will be settled for regularly through the communal authorities.

I expect the population to meet these necessities of war without resistance, and especially that there shall be no aggression against our troops, and that the supplies required shall be promptly furnished.

In this case I give every guarantee for the preservation of the city and the safety of its inhabitants.

If, however, as has unfortunately happened in other places, there are attacks upon our troops, firing upon our soldiers, fires or explosions of any sort, I shall be obliged to take the severest measures.

The General Commanding the Army Corps,

SIXT VON ARNIM.

The strongest thing so far was the series of demands made upon the city and Province. The city of Brussels has been given three days to hand over 50 million francs in coin or bills. The Germans also demand a tremendous supply of food to be furnished during the next three days. If the city fails to deliver any part of it, it must pay in coin at a rate equal to twice the market value of the supplies. The Province of Brabant must hand over, by the first of next month, 450 millions of francs—90 million dollars. When you consider that the total war indemnity imposed by Germany upon France in 1870 was only five milliards, the enormity of this appears. Upon one little province of a tiny country they are imposing a tax equal to one-tenth that imposed on the whole of France. How on earth they are ever to arrange to pay it, I cannot possibly see. I do not know what is to happen if they fail to make good, but I have no doubt that it will be something pretty dreadful.

This afternoon the Germans went into the Ministry of War and the Foreign Office, and searched through the archives. It must have been an entirely futile proceeding, for all papers of any interest were removed to Antwerp when the Government left. The higher officials who were still here were kept in the buildings to witness the search—a needless humiliation. There is talk now of a search of the British Legation, but we have heard nothing of it and expect that will not be done without asking our permission first.

Brussels, August 22, 1914.—Another day with much to do and no great results.

This morning, at 7 o'clock, General von Jarotzky arrived at the Legation and was all smiles. It appears that my action, in making known my displeasure at his behaviour and that of his staff, had a good effect. We have heard, from several sources, that he blew up everybody in sight yesterday afternoon when he came out from the Burgomaster's office and learned that I had departed in bad temper. He knows that nobody dares to oppose his acts or views, but just the same he gave them fits for not having made me stay and attend to my case. Be that as it may, he appeared with his Chief of Staff, and sent up a message that brought the Minister down in his pajamas and dressing gown. He expressed great regret for the "misunderstanding" of yesterday evening, and assured the Minister that there would be no further cause for complaint on our part. He had in his hand the telegram which we had sent him the evening before—the very same telegram which we had been trying to get

Types of Belgian cavalrymen

Civilian volunteers going out to dig trenches about Antwerp

The King of the Belgians (centre) giving orders in the field during the battle at H——

Sämtliche Militärbehörden werden hiermit dringend ersucht, den
Vorzeiger dieses

Passierscheines,

Herrn Hugh S. Gibson, Sekretär der
Gesandtschaft der Vereinigten Staaten von
Amerika in Brüssel in einem Auto mit
Chauffeur unbehindert nach Antwerpen
und zurück reisen zu lassen.

Herr Gibson will sich wegen des Schutzes
der deutschen Reichsangehörigen mit dem ame-
rikanischen Generalkonsul in Antwerpen
in Verbindung setzen.

Brüssel, 24. August 1914.
Deutsches Gouvernement.

von Jarotzky

Generalmajor und Gouverneur

Pass issued by General von Jarotzky, the first German commander in Brussels, to enable Mr. Gibson
to go through the lines to Antwerp.

off ever since the German occupation of the city. He had signed each page of the message, and had affixed his stamp with an order that it be immediately transmitted. He explained to the Minister that the best thing to do was for him to take it in person to the office of the Director of the Bureau of Telegraphs, who had already received instructions on the subject.

The servants were thrown into a perfect panic by the arrival of the *Généraux*. It took some argument to convince them that the Germans would hardly need to send two generals to take them into custody, even if they had any reason to desire them as prisoners.

About ten o'clock I was starting to go down to the telegraph office, to send the messages, when the Spanish Minister drove up in his big green car with the Spanish flag flying at the fore. We told him our story, whereupon he announced that he also had telegrams to send and that he would go with us. We drove in state to the telegraph office, and found that the entrance which had been indicated to us was the alley through which the mail wagons drive in the good days when there are any. Before an admiring crowd, we descended and made our way among Prussian troopers through the noisome alley to a small side door, where we were stopped by a sentry who stuck a bayonet in our general direction and said we could go no further. I was immediately thrust into the foreground as the brilliant German scholar; and, limbering up my heavy German artillery, I attacked him. The sentry blanched, but stood his ground. An officer came up as reinforcements, but was also limited to the German tongue; so I had to keep it up, with two full-grown Ministers

behind me thinking up impossible things to be translated into the hopeless tongue. The officer, who was
a genial soul, announced as though there were no use
ever again to appear at that particular place, that
the instruments had all been removed, and that there
was absolutely no way of sending any messages—no
matter from whom they came. We told him that we
had come at the special request of the General himself.
He replied that that made no difference whatever; that
if there were no wires and no instruments, there was
no possible way of sending the messages. After three
or four repetitions, the Minister and I began to understand that there was no use haggling about it; but
the Spanish Minister was not so lightly to be turned
aside and took up the cudgels, himself bursting into
the German language. He stood his ground valiantly
in the face of a volley of long words, but he did not
get any forrader. Prince Ernst de Ligne came in with
a permit from the General to send his messages, and
joined forces with the Spanish Minister; but the poor
officer could only shrug his shoulders and smile and
repeat what he had already said a score of times.
Mr. Whitlock and I began to laugh, and had a hard
time to control ourselves. Finally we prevailed upon
them to return to the Hôtel de Ville. The Minister was beginning to get even madder than he
was yesterday, when I got back with my story of the
way I had spent the afternoon, going from one wild
goose chase to another. We got the Burgomaster in
his private office and placed our troubles before him.
He understood the importance of the matter and sent
for the General. He appeared in short order, clicked

his heels, and inquired whether we had come in regard
to the matter of telegrams. The old fox knew perfectly
well that we had, and was ready for us. We had come
to the conclusion—which I had reached yesterday
afternoon and held all by myself—that the old man
was jockeying.

He listened to what we had to say, and then said
that there was no means of communication with the
outside world; that he had just learned it a few minutes
before. It is hardly necessary to say that he had been
fully posted from the minute he set foot in the town.
The Spanish Minister was rather sarcastic about his
opinion of a General who would venture to occupy
a capital without being in possession of means of
telegraphic communication. The old soldier was in
no mood for argument on abstract questions, and was
playing for too big stakes to stop and dicker, so he
passed this over lightly and suggested that we go
back and discuss with the Director-General of Telegraphs the possibilities of reëstablishing communications. Then the Spanish Minister let loose on him,
and announced that it was not consistent with the
dignity of representatives of World Powers to spend
their time standing in back alleys disputing with
soldiers who barred the way and refused to honour
the instructions of their General. He threw in hot
shot until the effect told. He said plainly that the
General was full of fair words and promises and agreed
to anything that was asked of him, but that when
we went to do the things he had authorised, we were
baffled by subordinates that took it upon themselves
to disregard these orders—the intimation being cleverly

conveyed that their action might not be unconnected with instructions from above. The old man then dropped his bluff, and asked what we wanted. We asked that he send for the Director-General, and give him, in our presence, the instructions and authorisation necessary to enable him to reëstablish communication with the outside world, and instruct him to receive and send all official messages for the Legations of neutral Powers. There was no way out, short of flatly refusing to give us our right to communicate with our governments, so the Director-General was sent for and the Burgomaster wrote out, at our dictation, the most general and comprehensive orders to meet our wishes in all matters of official business. The General signed the order and instructed the Director-General to go ahead.

The Director-General was a poor soul who could see nothing but technical difficulties in everything that was proposed. He reluctantly agreed to everything that he was told to do, and there is no telling when our stories will get off. He told us that when the Germans had occupied the telegraph bureau, instead of simply disconnecting the instruments and placing a man there to see that communication was not reëstablished, the officer in command had battered down the door leading to the roof and had slashed all the wires with his sabre. As there were three or four hundred wires leading out of the office, it will be a tremendous job to get them all together again.

We also took occasion to arrange for the issuance of *sauf conduits* for all the members of the Legations and for such members of the foreign colonies under our protection as we care to vouch for. Food is getting very scarce because of the enormous demands of the Germans, and we told von Jarotzky that we should expect that he make arrangements to see that our colonies should not suffer from the requisitions—that ample food be reserved to keep them all as long as it might be found necessary for them to stay here. He agreed to this, but I don't see just how he is to arrange it in practice. There are about fifty thousand men camping within a few miles of Brussels, and another Army Corps is now marching in. The food for all the people must be supplied by the city—all importations from the outside world have been suspended for days. It is a pretty bad situation, and it will probably get a great deal worse before long. I don't know whether we shall get down to eating horse and dog, but it is not altogether improbable. That is one of these things that it is interesting to read about afterward.

We spent nearly two hours at the Hôtel de Ville, and got in a good deal of talk that will be of service to all sorts of people. When we got back, we found the chancery full of people who were waiting for us to tell them just how they could send telegrams and letters, and get passports and permits to pass through the lines in all possible directions. Before leaving I had dictated a bulletin which was posted in the hallway, stating that there were no communications with the outside world by rail, telegraph or post, and that no *laisser-passers* would be granted by the authorities until conditions had changed, and that the Legation could not issue any sort of papers which would enable people to leave in safety.

About four o'clock, McCutcheon, Irwin and Cobb breezed in, looking like a lot of tramps. Several days ago they had sailed blissfully away to Louvain in a taxi, which they had picked up in front of the hotel. When they got there, they got out and started to walk about to see what was going on, when, before they could realise what was happening, they found themselves in the midst of a Belgian retreat, hard-pressed by a German advance. They were caught between the two, and escaped with their lives by flattening themselves up against the side of a house while the firing continued. When the row was over, they were left high and dry with no taxi—of course it had been seized by the retreating troops—and with no papers to justify their presence in Louvain at such a time. They decided that the best thing to do was to go straight to the German headquarters and report. They were received well enough, and told to lodge themselves as best they could and stay indoors until it was decided what was to be done with them. They were told that they might be kept prison-ers here, or even sent to Berlin, but that no harm would come to them if they behaved themselves. The order had gone out that if a single shot was fired at the German troops, from the window of any house, everybody in the house was to be immediately taken out and shot. Not wishing to risk any such unpleasant end, they rented all the front rooms of a house and spread themselves through all the rooms, so that they could be sure that nobody did any slaughtering from their house. They were there for three days, and were told to-day that they might

A Belgian machine gun battery drawn by dogs

Types of Belgian infantrymen

King Albert and General von Emmich who commanded the German troops at Liège. Taken in 1913 when General von Emmich came to Liège on a visit of courtesy.

take themselves hence. They came back to Brussels in the same clothes that they had worn for the past three days, unshaven and dirty. When they drove up to the front door this afternoon, they were nearly refused admittance as being too altogether disreputable.

This evening, when I went to see my old friend the General, just before dinner, he told me that he had had news of a great battle near Metz, in which the French army had been cut off and practically destroyed, with a loss of 45,000 prisoners. It sounds about as probable as some of the other yarns. In view of the fact that my friend had no telegraphic communication, I was curious to know where he got his information, but my gentle queries did not bring forth any news on that point.

The Germans now expect to establish themselves for some time here in Brussels. They are going to occupy the various governmental departments, and it is quite possible that for some time we shall have to deal exclusively with them. The Government to which we are accredited has faded away, and we are left here with a condition and not a theory. We shall have to deal with the condition, and I am not at all sure that the condition will not require some pretty active dealing with. Functionaries are to be brought from Berlin to administer the various departments, so that it is evidently expected that the occupation is not to be of a temporary character.

Later.—After writing the foregoing, I went upstairs and listened to some of the tales of the four people

who were tied up at Louvain. Now that they are safely
out of it, they can see the funny side of it, but it was
certainly pretty dangerous while it lasted. Monsieur
de Leval is overcome with admiration for their *sang-
froid*, and marvels at the race of men we breed.

They seem to have made themselves solid with the
Germans before they had been there long; it would
be hard for anybody to resist that crowd any length of
time. Of course they never saw their taxi again after
getting out to scout for the battle, and whenever the
Major who had the duty of keeping them under surveil-
lance came to take a look at them, Cobb would work up
a sob-shaken voice and plead for liberty and permission
to return to Brussels. He was always at some pains
to explain that it was not his life he was worrying
about, but the haunting thought of that taxi running
up at the rate of fifty centimes every three minutes.
After a while he got the Major's funny bone located,
and then all was well. He so completely got into the
officer's good graces that he promised to send us word
that they were safe and well,—and then failed to do
so.

While the Germans occupied the city, all inhabitants
were required to be indoors by eight o'clock; a light
had to be kept in every window, and the blinds left
open, so that any one moving could be clearly seen from
the street. The windows themselves were to be closed.
Dosch said he woke up about four o'clock one morning with his head splitting; the lamp was smoking and
the air vile with smoke and smell. He decided he
would prefer to be shot than die of headache, so
deliberately got up and opened his window. The

story loses its point by the fact that after violating this strict rule, he was not taken out and shot.

They said it was really pretty dreadful. From their window, they saw, every little while, a group of soldiers lead some poor frightened Belgian to a little café across the street; several officers were sitting at one of the tables on the sidewalk, holding a sort of drum-head court martial. While they were examining the case, a squad would be marched around behind the railroad station. A few minutes later the prisoner would be marched around by another way, and in a few minutes there would be a volley and the troops would be marched back to their post; then, after a little while, a stretcher would be brought out with a body in civilian clothes, a cloth over the face. Some of the prisoners were women, and there were screams before the shots were fired. It must have been a dreadful ordeal to go through.

Brussels, August 27, 1914.—The day after my last entry I started on a trip to Antwerp, got through the lines and managed to wriggle back into Brussels last night after reëstablishing telegraph communication with the Department and having a number of other things happen to and around about me.

All I can remember now of the 23d is that it was a Sunday, and that we could hear cannonading all day long from the east. It was hard to tell just where it came from, but it was probably from the direction of Wavre and Namur. It was drizzly all day. The German troops continued to pour through the city. From time to time, during the last few days, their march has been interrupted for a couple of hours at a time, apparently as a result of a determined attempt on the part of the French and English to stop the steady flow of troops toward the French frontier. Each time we could hear the booming of the cannon, the deep voices of the German guns and the sharp, dry bark of the French. At night we have seen the searchlights looking for the enemy or flashing signals. Despite the nearness of all this fighting and the sight of the wounded being brought in, the streets barred off to keep the noisy traffic away from the hospitals, and all the other signs of war, it has still been hard to realize that it was so near us.

Our little German General, von Jarotzky, has kept clicking his heels together and promising us anything we chose to ask. We have run around day after day with our telegrams, and not one has got farther than the Hôtel de Ville. Being naturally somewhat touchy, we got tired of this after a few days, and decided that the only way to get any news to Washington was for me to go to Antwerp and get into direct communication over the cable from there. We got our telegrams ready and made a last try on the General Monday morning. He was still effusively agreeable and assured us that he had determined to place a military field wire at our disposal so that we could communicate with Washington via Berlin. Our previous experiences had made us suspicious, so it was decided that while depositing our messages here, I would make a try at getting through the lines and send whatever I thought best from Antwerp or any other place I could reach. We told the General frankly what we intended to do, and he was all smiles and anxiety to please. At our request he had an imposing passport made out for me, signed with his hand and authorized with his seal. The Burgomaster wrote out an equally good letter for us when we reached the Belgian lines. Providence was to take care of us while we were between the lines, and, just to make it unanimous, He did.

We wanted to get away during the morning, but one thing after another came up and I was kept on the jump. We had to stop and worry about our newspaper correspondents, who have wandered off again. Morgan came sauntering in during the morning and announced that he and Davis had set out on foot to see whether there was any fighting near Hal; they had fallen in with some German forces advancing toward Mons. After satisfying themselves that there was nothing going on at Hal or Enghien, Morgan decided that he had had enough walking for one day, and was for coming home. Davis felt that they were too near the front to give up, and with a Sherlock Holmes sagacity announced that if they stuck to these German troops, they would succeed in locating the French and British armies. Morgan thought this so probable that he was all for coming back, and left Davis tramping along behind an ammunition wagon in search of adventure. He found it.

After getting out of their trouble at Louvain, McCutcheon, Cobb and Lewis set forth on another adventure. There are, of course, no motor cars or carriages to be had for love or money, so they invested in a couple of aged bicycles and a donkey cart. Cobb, who weighs far above standard, perched gracefully on top of the donkey cart, and the other two pedalled alongside on their wheels. They must have been a funny outfit, and at last accounts were getting along in good style. The air is filled with nervousness, however, and there is a constantly increasing list of people who are being thrown into jail, or shot as spies, and there is little time for careful and painstaking trials for wanderers who are picked up inside the lines of the fighting armies and are unable to render a convincing account of themselves. I shall be rather uncomfortable about them until they reappear.

While we were waiting for the final formalities for our trip to be accomplished, I invested in a wrist watch and goggles. We also bought a fuzzy animal like a Teddy bear, about three inches high, and tied him on the radiator as a mascot. He made a hit with all hands and got a valuable grin from several forbidding-looking Germans. We had signs on the car fore and aft, marking it as the car of the American Legation, the signs being in both French and German. As we were the first to try to make the trip, we thought it up to us to neglect nothing that would help to get us through without any unpleasant shooting or bayoneting.

After formally filing all our telegrams with the German General, Blount and I got under way at half-past two. We pulled out through the northern end of the city, toward Vilvorde. There were German troops and supply trains all along the road, but we were not stopped until we got about half way to Vilvorde. Then we heard a loud roar from a field of cabbages we were passing and, looking around, discovered what

to the table, filling up all that part of the room. As we were finishing dinner, several Ministers of State came in to say that the Prime Minister wanted me to come to meet him and the Cabinet Council which was being held—just to assure them that all was well with their families and to tell them, in the bargain, anything that I felt I properly could. However, I had my real work ahead of me—getting off my telegrams to Washington. I tore myself away from the crowd and, joining Sherman, who was waiting for me in the hall, I made for the Consulate-General. The Consul-General was already there, anxious to hear the news. I had to get before the Department all the news I could, and as comprehensive a statement as possible of everything that had happened since communications had been cut. I pounded away until after eleven, and got off a fat bundle of cables, which Sherman took to the office for me. I then made for the Grand Hôtel, where the Cabinet Council was waiting for me.

I have never been through a more moving time than the hour and a half I spent with them. It was hard to keep from bursting out and telling them everything that I knew would interest them. I had bound myself with no promises before I left about telling of the situation, but none the less I felt bound not to do it. I was able to tell them a great deal that was of comfort to them, and that could give no ground for objection if the Germans were to know of it, and, on these subjects, I gave them all they wanted. After telling them all I could about their families and friends, I let them ask questions and did my best to answer those that

I could. The first thing they wanted to know was how the Germans had behaved in the town. The answer I gave them was satisfactory. Then they wanted to know whether the Royal Palace had been respected, or whether the German flag was flying over it; also whether the Belgian flag still flew on the Hôtel de Ville. Their pride in their old town was touching, and when they heard that no harm had as yet been done it, you would have thought that they were hearing good news of friends they had lost. Then they started in and told me all the news they had from outside sources—bits of information which had reached them indirectly via Holland, and the reports of their military authorities. We have never had such complete information given us—enough to justify the trip even if I had not restored communication with the Department.

We stayed on and talked until nearly half-past twelve, when I got up and insisted on leaving; perhaps it is just as well. They did not want to break up the party, but when I insisted, they also made up their minds to call it a day's work and quit.

We brought van der Elst back to the hotel, and with his influence ran our car into the Gendarmerie next door. Then to bed.

Blount and I had a huge room on the third floor front. We had just got into bed and were settling down to a good night's rest when there was an explosion, the like of which I have never heard before, and we sat up and paid strict attention. We were greatly interested, but took it calmly, knowing that the forts were nearly four miles out of town and that they could bang away as long as they liked without doing more than spoil our night's sleep. There were eight of these explosions at short intervals, and then as they stopped there was a sharp *purr* like the distant rattle of a machine gun. As that died down, the chimes of the Cathedral—the sweetest carillon I have ever heard—sounded one o'clock. We thought that the Germans must have tried an advance under cover of a bombardment, and retired as soon as they saw that the forts were vigilant and not to be taken by surprise. We did not even get out of bed. About five minutes later we heard footsteps on the roof and the voice of a woman in a window across the street, asking some one on the sidewalk below whether it was safe to go back to bed. I got out and took a look into the street. There were a lot of people there talking and gesticulating, but nothing of enough interest to keep two tired men from their night's sleep, so we climbed back into bed and stayed until morning.

Blount called me at what seemed an unreasonably early hour and said we should be up and about our day's work. When we were both dressed, we found that we had made a bad guess, when he looked at his watch and discovered that it was only a quarter to seven. Being up, however, we decided to go down and get our breakfast.

When we got down we found everybody else stirring, and it took us several minutes to get it through our heads that we had been through more excitement than we wotted of. Those distant explosions that we had taken so calmly were bombs dropped from a Zeppelin which had sailed over the city and dropped death and destruction in its path. The first bomb fell less than two hundred yards from where we slept—no wonder that we were rocked in our beds! After a little breakfast we sallied forth.

The first bomb was in a little street around the corner from the hotel, and had fallen into a narrow four-story house, which had been blown into bits. When the bomb burst, it not only tore a fine hole in the immediate vicinity, but hurled its pieces several hundred yards. All the windows for at least two hundred or three hundred feet were smashed into little bits. The fronts of all the surrounding houses were pierced with hundreds of holes, large and small. The street itself was filled with débris and was impassable. From this place we went to the other points where bombs had fallen. As we afterward learned, ten people were killed outright; a number have since died of their injuries and a lot more are injured, and some of these may die. A number of houses were completely wrecked and a great many will have to be torn down. Army officers were amazed at the terrific force of the explosions. The last bomb dropped as the Zeppelin passed over our heads fell in the centre of a large square—la Place du Poids Publique. It tore a hole in the cobblestone pavement, some twenty feet square and four or five feet deep. Every window in the square was smashed to bits. The fronts of the houses were riddled with holes, and everybody had been obliged to move out, as many of the houses were expected to fall at any time. The Dutch Minister's house was near one of the smaller bombs and was damaged slightly. Every window was smashed. All the crockery and china are gone; mirrors

Boy Scouts at Belgian headquarters, Lierre

Reading from left to right: a Belgian Staff Officer, Colonel Fairholme, Colonel DuCane and Captain Ferguson. (Malines Cathedral in the background)

List of the civilians killed by the Germans at Tamines on
August 20, 1914

in tiny fragments; and the Minister somewhat startled. Not far away was Faura, the First Secretary of the Spanish Legation. His wife had been worried sick for fear of bombardment, and he had succeeded only the day before in prevailing upon her to go to England with their large family of children. Another bomb fell not far from the houses of the Consul-General and the Vice-Consul-General, and they were not at all pleased. The windows on one side of our hotel were also smashed.

We learned that the Zeppelin had sailed over the town not more than five hundred feet above us; the motor was stopped some little distance away and she slid along in perfect silence and with her lights out. It would be a comfort to say just what one thinks about the whole business. The *purr* of machine guns that we heard after the explosion of the last bomb was the starting of the motor, which carried our visitor out of range of the guns which were trundled out to attack her. Preparations were being made to receive such a visit, but they had not been completed; had she come a day or two later, she would have met a warm reception. The line of march was straight across the town, on a line from the General Staff, the Palace where the Queen was staying with the royal children, the military hospital of Ste. Elisabeth, filled with wounded, the Bourse, and some other buildings. It looks very much as though the idea had been to drop one of the bombs on the Palace. The Palace itself was missed by a narrow margin, but large pieces of the bomb were picked up on the roof and shown me later in the day by Inglebleek, the King's Secretary. The room at the

General Staff, where I had been until half an hour before the explosion, was a pretty ruin, and it was just as well for us that we left when we did. It was a fine, big room, with a glass dome skylight over the big round table where we were sitting. This came in with a crash and was in powder all over the place. Next time I sit under a glass skylight in Antwerp, I shall have a guard outside with an eye out for Zeppelins.

If the idea of this charming performance was to inspire terror, it was a complete failure. The people of the town, far from yielding to fear, are devoting all their energies to anger. They are furious at the idea of killing their King and Queen. There is no telling when the performance will be repeated, but there is a chance that next time the balloon man will get a warmer reception.

In the morning I went around and called at the Foreign Office, which is established in a handsome building that belonged to one of the municipal administrations. The Minister for Foreign Affairs took me into his office and summoned all hands to hear any news I could give them of their families and friends. I also took notes of names and addresses of people in Brussels who were to be told that their own people in Antwerp were safe and well. I had been doing that steadily from the minute we set foot in the hotel the night before, and when I got back here, I had my pockets bulging with innocent messages. Now comes the merry task of getting them around.

At the hotel we were besieged with invitations to lunch and dine with all our friends. They were not only glad to see somebody from the outside world, but could not get over the sporting side of our trip, and patted us on the back until they made us uncomfortable. Everybody in Antwerp looked upon the trip as a great exploit, and exuded admiration. I fully expected to get a Carnegie medal before I got away. And it sounded so funny coming from a lot of Belgian officers who had for the last few weeks been going through the most harrowing experiences, with their lives in danger every minute, and even now with a perfectly good chance of being killed before the war is over. They seem to take that as a matter of course, but look upon our performance as in some way different and superior. People are funny things.

I stopped at the Palace to sign the King's book, and ran into General Jungbluth, who was just starting off with the Queen. She came down the stairs and stopped just long enough to greet me, and then went her way; she is a brave little woman and deserves a better fate than she has had. Inglebleek, the King's Secretary, heard that I was there signing the book, and came out to see me. He said that the Queen was anxious I should see what had been done by the bombs of the night before. He wanted me to go right into the houses and see the horrid details. I did not want to do this, but there was no getting out of it under the circumstances.

We drove first to the Place du Poids Publique and went into one of the houses which had been partially wrecked by one of the smaller bombs. Everything in the place had been left as it was until the police magistrate could make his examination and report. We climbed to the first floor, and I shall never forget the horrible sight that awaited us. A poor policeman and his wife had been blown to fragments, and the pieces were all over the walls and ceiling. Blood was everywhere. Other details are too terrible even to think of. I could not stand any more than this one room. There were others which Inglebleek wanted to show me, but I could not think of it. And this was only one of a number of houses where peaceful men and women had been so brutally killed while they slept.

And where is the military advantage of this? If the bombs were dropped near the fortifications, it would be easy to understand, but in this instance it is hard to explain upon any ground, except the hope of terrifying the population to the point where they will demand that the Government surrender the town and the fortifications. Judging from the temper they were in yesterday at Antwerp, they are more likely to demand that the place be held at all costs rather than risk falling under the rule of a conqueror brutal enough to murder innocent people in their beds.

The Prime Minister told me that he had four sons in the army—all the children he has—and that he was prepared to give every one of them, and his own life and fortune, into the bargain, but that he was *not* prepared—and here he banged his fist down on the table and his eyes flashed—to admit for a minute the possibility of yielding to Germany. Everybody else is in the same state of mind. It is not hysterical. The war has been going on long enough, and they have had so many hard blows that the glamour and the fictitious attractiveness of the thing has gone, and they have settled down in deadly earnest to fight to the bitter end. There may not be one stone left upon another in Belgium when the Germans get through, but if these people keep up to their present level they will come through—what there is left of them—free.

Later in the afternoon I went to the Foreign Office and let them read to me the records of the commission which is investigating the alleged German atrocities. They are working in a calm and sane way and seem to be making the most earnest attempt to get at the true facts, no matter whether they prove or disprove the charges that have been made. It is wonderful to see the judicial way they can sit down in the midst of war and carnage and try to make a fair inquiry on a matter of this sort. If one one-thousandth part of the charges are proven to be true. . . .

The rest of the afternoon was spent seeing people who came in for news of Brussels, and who had messages to send home. I had had to tell the hotel people that I would be there from four to seven to see people, and that the rest of the time I must have free for my own work. They came in swarms; all the diplomats, the Cabinet Ministers, and the Ministers of State, army officers, and other officials—a perfect mob. I had a package of cards on which I noted names and addresses and the messages which were to be delivered. These messages have been sent out to-day, after being submitted to the military authorities, some of them in writing and some by word of mouth, and if they have afforded one-tenth the comfort that I hope, the sum total of misery in this town has been reduced a good deal this day.

Colonel Fairholme left for the front, with the King, early in the morning, and was with him during the battle at Malines. He thought we were going back during the day, as I had told him the evening before. About noon he called up from the telephone and told Sir Francis that under no circumstances was I to be allowed to start, as the town was being bombarded with heavy siege pieces and all traffic was absolutely stopped; that we could not only not get by, but that any part of the trip by the regular road was extremely dangerous. I was just as glad that we had decided to stay over. The Colonel stayed out all that night and had not returned to Antwerp when we left yesterday. During the morning he called up again and asked about us, again advising against our starting. Pretty decent of a man who has as much to think of as he had to be worrying about us enough to take time to telephone us as to the dangers of the road.

During the evening bad news came in from France, and everybody was down in the mouth. The French Minister came in and told me what he had received. Everybody was plainly worried, and altogether things looked pretty dismal. We sat around a little while and then decided for a good night's sleep.

To make sure of offering no unnecessary chances for Mr. Zeppelin the authorities had ordered all the lights on the streets put out at eight o'clock. It was dark as midnight and there was no use in thinking of venturing out into the town. The Cathedral clock was stopped and the carillon turned off for the first time in heaven only knows how many years. It was a city of the dead. Guns were posted in the streets ready for instant use in case the airship should put in another appearance. As a result of this and the searchlights

that played upon the sky all night, our friend the enemy
did not appear. Some people know when they have
had enough.

Yesterday morning I looked out of my window at
the Cathedral clock, and saw that it was twenty-five
minutes to ten. I tumbled through my tub, and
rushed downstairs to get through my morning's work,
only to find that it was half-past six. I had forgotten
that the Cathedral clock had been stopped.

It was just as well that I was up early, however, for
there was plenty to be done. I found a lot of telegrams
waiting for me at the Consulate, and had to get off
another string of them. Then an orderly held me up
on the street to tell me that the King's Secretary was
hunting for me all over the place, and that I was
wanted at the Palace. When I got there, he had
started off on another hunt for me. He finally got
me at the hotel, and kept me for half an hour.

By the time that I got through with him, there was
word that the Minister of Foreign Affairs wanted to
see me, so I made a bee-line over there. Then there
was another call to the Consulate to answer some more
telegrams. After attending to various matters at the
Palace, the Ministry for Foreign Affairs, the Consulate
General, and seeing a few more people at the hotel,
the morning was gone and it was time for lunch and
a quick get-away.

All hands came out and bade us farewell. You
would have thought we were on our way to Heaven,
except for the fact that they urged us to come back.

As we could hear the cannonading, we decided that
we would avoid the Malines road and would try to

skirt around the zone of trouble and work our way into Brussels from the west. We got ferried across the Scheldt on a terrible tub of a steamer that looked as though she would go down under the weight of the military automobiles that she had to get across in order to take ammunition to the front. We all got away in a bunch from the other side, but we drew ahead of them as we had not such a heavy load; and within three-quarters of an hour we were outside the Belgian lines. Van der Elst had secured for us a most imposing *laisser-passer*, which took us through with practically no trouble except that it was so impressive that we were held at each barricade while all the men on duty took turns reading it. The only ticklish part of the trip to the Belgian outposts was working our way through the villages which had been mined in anticipation of a German invasion. It is bad enough working one's way through there in a motor with everybody helping you to keep out of harm's way, but it must be a trifle worse to do it in a mass with a man on a hill a little way off waiting for you to come up to the signal post so that he can touch a button and send you in small pieces into the next world.

We struck out through St. Nicholas, Hamme, Termonde and Assche, and got into Brussels from the west without mishap. We have got quite used to having people poke bayonets in our faces and brandish revolvers at us, so the latter part of the trip with only that to contend with seemed quiet and almost boring.

On the road in from Assche, we passed near Eppeghem and Vilvorde, where the fighting had been going on for a couple of days. After news had been received in Antwerp of the defeat of the French and English at Mons and Charleroi, the Belgians were ordered to fall back on Antwerp and had left these little villages to be occupied by the Germans. As they occupied them, they had set them afire and the flames were raging as we came by. They were quaint little towns, and had excited our admiration two days before when we had gone through—despite the fact that we had other things on our minds beside admiring the beauties of architecture. Now they are gone.

The Germans gave us no trouble, and we got back to the Legation by a little before five. Everyone poured out to meet us, and greeted us as prodigal sons. When we had not come back the day before, they had about made up their minds that something dreadful had happened to us, and the rejoicing over our return was consequently much greater than if we had not whetted their imaginations just a little.

I found that the situation in Brussels had undergone big changes while I was away. General von Jarotzky had been replaced by General von Lüttwitz, who is an administrator, and has been sent to put things in running order again. There was no inkling of this change when I left, and I was a good deal surprised. Guns have been placed at various strategic points commanding the town, and the Germans are ready for anything. The telephone wire they had put through the town to connect the two stations and headquarters was cut day before yesterday by some cheerful idiot who probably thought he was doing something good for his country. The military authorities thereupon announced that if anything of the sort was done again they would lay waste the quarter of the town where the act was committed.

Some of the subordinate officers have since told us that von Jarotzky was a fighting general, and had no business staying in a post requiring administrative ability. The new man is cut out particularly for this sort of work, and is going to start a regular German administration. Functionaries are being brought from Berlin to take things over, and in a short time we shall, to all intents and purposes, be living in a German city. The first trains ran to-day in a halting fashion to Liège and the German frontier. Perhaps we shall have a newspaper.

Most distressing news has come through from Tamines. I had a long talk to-day with a trustworthy man from there, and his story was enough to make one's blood run cold. He says that on the evening of the twenty-first the Germans entered the village after a brush with French troops which were still in the neighbourhood. Infuriated by the resistance offered to their advance, they proceeded to vent their rage on the town. They shot down a lot of villagers, and arrested many more. A great many escaped to the country. A lot of houses were first sacked, and then burned. The orgy continued during the night, and through the next day. On the evening of the twenty-second, something over four hundred men were collected near the church and lined up to be shot. The work was done for a time by a firing squad which fired into the crowd with more or less system, but this was too slow, and finally a rapid-fire gun was brought out and turned loose. Of course, a great many were not killed outright and lay groaning among the dead. Now and then a German would put one out of his misery by a bayonet thrust. Others settled their own troubles by rolling themselves into the nearby river. Altogether over six hundred people were shot down, but it is hard to get any exact figures yet. After the shooting was over, other civilians were brought out and compelled to bury the dead. My informant says that some of the scenes attending this duty were quite as poignant as the shooting itself, for some buried their own fathers and brothers. One man about to be thrown into the trench was found to be still alive, but the German doctor, after a cursory examination, ordered him buried with the rest. The man had enough life left in him to raise his hand in appeal but the doctor shrugged his shoulder and repeated his order. There were many incidents, most of them horrible. The man who told the story seemed still dazed and spoke quietly, with few adjectives and little emphasis on anything he said. It was a bare recital of facts, and far more moving than if he had striven for effect.

Davis got back yesterday from his trip to the front, and we learned that he had been through a perfectly good experience that will look well when he comes to writing it up, but one that gave him little satisfaction while it was in progress. He started off to follow the German army in the hope of locating the English. After leaving Hal, some bright young German officer decided that he was a suspicious-looking character, and ought to be shot as an English spy. As a preliminary, they arrested him and locked him up. Then

the war was called off while the jury sat on his case.
One of the officers thought it would be a superfluous
effort to go through the form of trying him, but that
they should shoot him without further to do. They
began considering his case at eleven in the morning,
and kept it up until midnight. He was given pretty clearly to understand that his chances were slim, and that
the usual fate of spies awaited him. He argued at
length, and apparently his arguments had some effect,
for at three o'clock in the morning he was routed out
and told to hit the road toward Brussels. He was
ordered to keep religiously to the main road all the
way back, on pain of being shot on sight, and to report
at headquarters here immediately on his arrival. By
this time he was perfectly willing to do exactly what
was demanded by those in authority, and made a bee-
line back here on foot. He turned up at the Legation
yesterday morning, footsore and weary, and looking
like a tramp, and told his story to an admiring audience.
I was still away on my little jaunt, and did not get
it at first hand. The Minister took him down to call on
the General, and got them to understand that Richard
Harding Davis was not an English spy, but, on the
contrary, probably the greatest writer that ever lived,
not excepting Shakespeare or Milton. The General
said he had read some of his short stories, and that he
would not have him shot. Just the same, he was not
keen about having him follow the operations. He is
now ordered to remain in this immediate neighbourhood until further orders. To-day he had several
interviews with the General in an attempt to get
permission to leave the country, but had no luck.

The last we saw of Davis, he came in late this afternoon to tell us that he did not know what to do next. He said that he had been through six wars, but that he had never been so scared as he was at that time. If he is allowed to get out of Belgium, I think that he will not darken the door of General von Lüttwitz for some time to come.

I was surprised to learn that Hans von Herwarth, who used to be military attaché in Washington, and whom I knew very well, is here as Adjutant to our new Governor. I have not yet had time to get over to see him, but shall try to do so to-morrow. I am glad to have somebody like that here to do business with. He is a real white man, and I anticipate a much better time with him than with any other officer they could send here in that capacity.

Baron Capelle came in late this afternoon to tell me that the Germans were bringing in a lot of priests on carts filled with cows and pigs, and were planning to hold them as hostages. One of them had called out and asked him to notify us that Monseigneur de Becker, Rector of the American College at Louvain, was among these prisoners. He is the priest I went to see when I was in Louvain ten days ago. I had told him he was perfectly safe, and scoffed at his fears.

The Minister was out when this news came, but I sallied forth and tried to locate the Monseigneur. He was not to be found anywhere. When I got back to the Legation, both the Minister and Villalobar were here and I told them all about what had happened. The people of the town were getting excited over the treatment that was being meted out to their priests, and it was in a fair way to result in serious trouble. Both Ministers made for the Ministry for Foreign Affairs, where the German Government is established, and before they left, had secured orders for the release of all the hostages. A lot of these terrible things are done by subordinate officers, and the people at the top seem only too anxious to learn of such affairs and do what they can to remedy them. The day has been dreadful with stories of suffering and murder and pillage.

Not only are we cut off from communication with the outside world, but a lot of the ordinary conveniences of life have already disappeared. We have no newspapers, no trams, no taxis, no telephones. Milk is no longer to be had, and within a day or two we shall have no butter or eggs. Then it will begin to look like a real siege. In a day or so I am to have a list of Jarotzky's demands for supplies, so that I can cheer myself with thoughts of what our life is to be like.

There is bad news from Louvain. The reports we have received agree that there was some sort of trouble in the square before the Hôtel de Ville a day or two ago. Beyond that, no two reports are alike. The Germans say that the son of the Burgomaster shot down some staff officers who were talking together at dusk before the Hôtel de Ville. The only flaw in that story is that the Burgomaster has no son. Some Belgians say that two bodies of Germans who were drunk met in the dusk; that one body mistook the other for French, and opened fire. Other reliable people tell with convincing detail that the trouble was planned and started by the Germans in cold blood. However that may be, the affair ended in the town being set on fire, and civilians shot down in the streets as they tried to escape. According to the Germans themselves, the town is being wiped out of existence. The Cathedral, the Library, the University, and other public buildings have either been destroyed or have suffered severely. People have been shot by hundreds, and those not killed are being driven from the town. They are coming to Brussels by thousands, and the end is not yet. This evening the wife of the Minister of Fine Arts came in with the news that her mother, a woman of eighty-four, had been driven from her home at the point of the bayonet and forced to walk with a stream of refugees all the way to Tervueren, a distance of about twelve miles, before she could be put on a tram to her daughter's house. Two old priests have staggered into the ——— Legation more dead than alive after having been compelled to walk ahead of the German troops for miles as a sort of protecting screen. One of them is ill, and it is said that he may die as a result of what he has gone through.

August 28th.—After lunch Blount and I decided to go out to Louvain to learn for ourselves just how much truth there is in the stories we have heard, and see whether the American College is safe. We were going alone, but Pousette and Bulle, the Swedish and Mexican Chargés d'Affaires, were anxious to join us, so the four of us got away together and made good time as far as the first outpost this side of Louvain.

Here there was a small camp by a hospital, and the soldiers came out to examine our papers and warn us to go no farther, as there was fighting in the town. The road was black with frightened civilians carrying away small bundles from the ruins of their homes. Ahead was a great column of dull gray smoke which completely hid the city. We could hear the muffled sound of firing ahead. Down the little street which led to the town, we could see dozens of white flags which had been hung out of the windows in a childish hope of averting trouble.

We talked with the soldiers for some time in an effort to get some idea of what had really happened in the town. They seemed convinced that civilians had precipitated the whole business by firing upon the staff of a general who was parleying with the Burgomaster in the square before the Hôtel de Ville. They saw nothing themselves, and believe what they are told. Different members of the detachment had different stories to tell, including one that civilians had a machine gun installed on top of the Cathedral, and fired into the German troops, inflicting much damage. One of the men told us that his company had lost twenty-five men in the initial flurry. They were a depressed and nervous-looking crew, bitter against the civil population and cursing their ways with great earnestness. They were at some pains to impress upon us that all Belgians were *Schwein*, and that the people of Louvain were the lowest known form of the animal.

After talking the situation over with the officer in command, we decided to try getting around the town to the station by way of the ring of outer boulevards.

Entrance to the Rue de Diest, Louvain

The dead and the living. A Belgian civilian and a German soldier

We got through in good shape, being stopped a few times by soldiers and by little groups of frightened civilians who were cowering in the shelter of doorways, listening to the noise of fighting in the town, the steady crackle of machine guns, and the occasional explosions.

They were pathetic in their confidence that the United States was coming to save them. In some way word has traveled all over Belgium that we have entered the war on the side of Belgium, and they all seem to believe it. Nearly every group we talked to asked hopefully when our troops were coming, and when we answered that we were not involved, they asked wistfully if we didn't think we should be forced to come in later. A little boy of about eight, in a group that stopped us, asked me whether we were English, and when I told him what we were, he began jumping up and down, clapping his hands, and shouting:

Les Américains sont arrivés! Les Américains sont arrivés!

His father told him to be quiet, but he was perfectly happy and clung to the side of the car as long as we stayed, his eyes shining with joy, convinced that things were going to be all right somehow.

About half way around the ring of boulevards we came to burning houses. The outer side of the boulevard was a hundred feet or so from the houses, so the motor was safe, but it was pretty hot and the cinders were so thick that we had to put on our goggles. A lot of the houses were still burning, but most of them were nothing but blackened walls with smouldering timbers inside. Many of the front doors had been battered open in order to start the fires or to rout out the people who were in hiding.

We came to a German ammunition wagon, half upset against a tree, where it had been hurled when the horses had turned to run away. The tongue was broken and wrenched out. Near by were the two horses, dead and swollen until their legs stood out straight. Then we began to see more ghastly sights— poor civilians lying where they had been shot down as they ran—men and women—one old patriarch lying on his back in the sun, his great white beard nearly hiding his swollen face. All sorts of wreckage scattered over the street, hats and wooden shoes, German helmets, swords and saddles, bottles and all sorts of bundles which had been dropped and abandoned when the trouble began. For three-quarters of a mile the boulevard looked as though it had been swept by a cyclone. The Porte de Tirlemont had evidently been the scene of particularly bloody business. The telegraph and trolley wires were down; dead men and horses all over the square; the houses still burning. The broad road we had traveled when we went to Tirlemont was covered with wreckage and dead bodies.

Some bedraggled German soldiers came out from under the gate and examined our passes. They were nervous and unhappy and shook their heads gloomily over the horrors through which they were passing. They said they had had hardly a minute's sleep for the past three nights. Their eyes were bloodshot and they were almost too tired to talk. Some of them were drunk—in the sodden stage, when the effect begins to wear off. They told us we could proceed in safety as far as the station, where we would find the headquarters of the commanding officer. Here we could leave the motor and learn how far we could safely go. This crowd varied the wording a little by saying that the Belgians were all dogs and that these particular dogs were being driven out, as they should be, that all that part of town was being cleared of people, ordered to leave their homes and go to Brussels or some other town, so that the destruction of Louvain could proceed systematically. We thought at the time that they were exaggerating what was being done, but were enlightened before we had gone much farther.

We continued down the boulevard for a quarter of a mile or so till we came to the station. Sentries came out and looked through our passes again. We parked the motor with a number of German military cars in the square and set off on foot down the Rue de la Station, which we had admired so much when we had driven down its length, just ten days before.

The houses on both sides were either partially destroyed or smouldering. Soldiers were systematically removing what was to be found in the way of valuables, food, and wine, and then setting fire to the furniture and hangings. It was all most businesslike. The houses are substantial stone buildings, and fire will not spread from one to another. Therefore the procedure was to batter down the door of each house, clean out what was to be saved, then pile furniture and hangings in the middle of the room, set them afire, and move on to the next house.

It was pretty hot, but we made our way down the street, showing our passes every hundred feet or so to soldiers installed in comfortable armchairs, which they had dragged into the gutter from looted houses, till we came to a little crossing about half way to the Hôtel de Ville. Here we were stopped by a small detachment of soldiers, who told us that we could go no farther; that they were clearing civilians out of some houses a little farther down the street, and that there was likely to be firing at any time.

The officer in command spoke to us civilly and told us to stick close to him so that we could know just what we ought to do at any time. He was in charge of the destruction of this part of the town and had things moving along smartly. His men were firing some houses near by and he stood outside smoking a rank cigar and looking on gloomily.

We exchanged remarks with him in German for a few minutes, I limping along behind the more fluent Pousette and Bulle. Then I said something in an aside to Blount, and the officer broke into the conversation in perfectly good English. He turned out to be a volunteer officer from Hamburg, who had spent some thirty years in England and was completely at home in the language.

We then accomplished the formal introductions which are so necessary to Germans even at a time like this, and when we came to Bulle the officer burst into a rapid fire of questions, which ended in his proclaiming in rapture:

"Why, I knew your father in Hamburg and went to school with your Uncle So-and-so!"

Reminiscence went on as though we were about a dining table at home; minute inquiry was made into

Señor don German Bülle, Mexican Chargé d'Affaires in Belgium

German sentries

Posing for a picture to send home to Germany

the welfare and activities of the Bulle family from the cradle to the grave. On the strength of the respectability of Bulle's relatives we were then taken under the officer's wing and piloted by him through the rest of our visit.

From where we stood we could see down the street through the smoke, as far as the Hôtel de Ville. It was still standing, but the Cathedral across the street was badly damaged and smoke was rising in clouds from its roof. The business houses beyond were not to be seen; the smoke was too dense to tell how many of them were gone.

Machine guns were at work near by, and occasionally there was a loud explosion when the destructive work was helped with dynamite.

A number of the men about us were drunk and evidently had been in that state for some time. Our officer complained that they had had very little to eat for several days, but added glumly that there was plenty to drink.

A cart, heaped high with loot, driven by a fat Landsturmer and pulled by a tiny donkey, came creaking past us. One of our party pulled his kodak from his pocket and inquired of our guardian in English: "May I take a picture?"

His intent evidently escaped the German, who answered cordially:

"Certainly; go ahead. You will find some beautiful things over there on the corner in the house they are getting ready to burn."

We kept our faces under control, and he was too much occupied with his other troubles to notice that we did not avail of his kind permission to join in the pillage.

He was rabid against the Belgians and had an endless series of stories of atrocities they had committed —though he admitted that he had none of them at first hand. He took it as gospel, however, that they had fired upon the German troops in Louvain and laid themselves open to reprisals. To his thinking there is nothing bad enough for them, and his chief satisfaction seemed to consist in repeating to us over and over that he was going the limit. Orders had been issued to raze the town—"till not one stone was left on another," as he said.

Just to see what would happen I inquired about the provision of The Hague Conventions, prescribing that no collective penalty can be imposed for lawless acts of individuals. He dismissed that to his own satisfaction by remarking that:

"All Belgians are dogs, and all would do these things unless they are taught what will happen to them."

Convincing logic!

With a hard glint in his eye he told us the purpose of his work; he came back to it over and over, but the burden of what he had to say was something like this:

"We shall make this place a desert. We shall wipe it out so that it will be hard to find where Louvain used to stand. For generations people will come here to see what we have done, and it will teach them to respect Germany and to think twice before they resist her. Not one stone on another, I tell you—*kein Stein auf einander!*"

I agreed with him when he remarked that people would come here for generations to see what Germany had done—but he did not seem to follow my line of thought.

While we were talking about these things and the business of burning and looting was pursuing its orderly course, a rifle shot rang out near by. Instantly every soldier seized his rifle and stood waiting for an indication as to what would happen next. In a few seconds a group of soldiers rushed into a house about a hundred feet away. There was a sound of blows, as though a door was being beaten in; then a few shots, and the soldiers came out wiping the perspiration from their faces.

"Snipers!" said our guide, shaking his fist at the house. "We have gone through that sort of thing for three days and it is enough to drive us mad; fighting is easy in comparison, for then you know what you are doing." And then almost tearfully: "Here we are *so* helpless!"

While he was talking another shot rang out, and then there was a regular fusillade, which lasted for fifteen seconds or so; then an explosion.

Bulle stood not upon the order of his going, but ran for the station, calling back:

"I've had enough of this. Let's get out and go home."

Our friend, the officer, said Bulle was right, and that it would be the part of wisdom for us all to fall back to the station, where we would be near the car in case anything happened. He started off at a good pace, and as we were in no mood to argue we went meekly along in his wake. We overtook Bulle engaged in an altercation with a very drunken soldier, who wanted to see his papers and was insulting about it. Instead of taking the easy course and showing his papers Bulle was opening a debate on the subject, when we arrived and took a hand. Our officer waded into the soldier in a way that would have caused a mutiny in any other army, and the soldier, very drunk and sullen, retreated, muttering, to his armchair on the curb. We then moved on to the station.

Outside the station was a crowd of several hundred people, mostly women and children, being herded on to trains by soldiers, to be run out of the town. They seemed to be decently treated but were naturally in a pitiable state of terror. Just inside the gates of the freight yard were a couple of women telling their troubles to a group of officers and soldiers. They had both lost their husbands in the street-fighting, and were in a terrible state. The officers and men were gathered about them, evidently distressed by their trouble, and trying to comfort them. They had put the older woman in an armchair and were giving her a little brandy in a tea cup. And the same men may have been the ones who killed the husbands. . . .

We went on into the freight yards and were greeted by a number of officers with hopeful talk about a train coming from Brussels with food. We were given chairs and an orderly was despatched for a bottle of wine so that a drink could be given to Bulle, who said that after what he had been through he would appreciate a glass of something comforting.

We settled down and listened to the stories of the past few days. It was a story of clearing out civilians from a large part of the town; a systematic routing out of men from cellars and garrets, wholesale shootings, the generous use of machine guns, and the free application of the torch—the whole story enough to make one see red. And for our guidance it was impressed on us that this would make people *respect* Germany and think twice about resisting her.

Suddenly several shots rang out apparently from some ruins across the street and the whole place was instantly in an uproar. The lines of civilians were driven helter-skelter to cover—where, I don't know. The stands of arms in the freight yard were snatched up, and in less time than it takes to tell it, several hundred men were scattered behind any sort of shelter that offered, ready for the fray.

I took one quick look about and decided that the substantial freight station was the most attractive thing in sight. In no time I was inside, closely followed by my own crowd and a handful of soldiers. First, we lay down upon the platform, and then, when we got our bearings, rolled over on to the track among a lot of artillery horses that were tethered there.

Apparently a number of civilians, goaded to desperation by what they had seen, had banded together, knowing that they were as good as dead, and had determined to sell their lives as dearly as they could. They had gathered in the ruins of the houses fronting on the station and had opened up on us. There was a brisk interchange of shots, with an occasional tinkle of broken glass and a good deal of indiscriminate cursing by the soldiers, who had taken refuge with us.

The artillery horses did not welcome us very cordially and began to get restive in a way that made us debate whether we preferred staying up on the platform with a chance of being potted or staying under cover and being ingloriously trampled to death. A joint debate on this important question kept us occupied for several minutes. We finally compromised by fishing down a few boxes from the platform and erecting a barricade of sorts to protect us against any stray kicks.

As we sat in the undignified position imposed on us by circumstances, we exchanged various frivolous remarks, not because we felt particularly gay, but because we had to do something to keep ourselves interested and to keep our courage up. Bulle resented this, and raised his head to look at me reproachfully over the barricade, and say: "Don't talk like that; it is nothing short of tempting Providence."

After a time Blount and I decided to make a reconnaissance in force and see how the car was getting on. We crawled along the floor to a place from which we could see out into the square. The soldiers were flat on their stomachs behind a low wall that extended around the small circular park in the centre of the square, and behind any odd shelter they could find. The car lay in the line of fire but had not been struck. We were sufficiently pessimistic to be convinced that it would go up in smoke before the row was over, and took a good look at our shoes to see whether they would last through a walk back to Brussels.

Our officer came out from behind his barricade and showed us where the attacking force was concealed— at least he told us that they were there and we were willing to take his word for it without going across the street to make a first-hand investigation.

He tried to impress us with the black sinfulness of people who would fire upon the German troops, and called our particular attention to the proof now offered us that civilians had started the row by firing on German troops. According to the German story, which was the only one we had heard, civilians had been hunted down like rats in garrets and cellars and shot down in cold blood in the streets when they sought safety in flight. To my mind it was not surprising that men driven to desperation by seeing their friends and neighbours murdered in cold blood, should decide to do any harm possible to the enemy. Three days of the reign of terror that had been described to us was enough to account for anything, and the fact that civilians were firing now did not in any sense prove that they were guilty of starting the trouble. For all we could tell they may have started it or they may not, but firing by them three days after the row began was no proof to any one with the slightest sense of the value of evidence. On the other hand, the story freely told us by the Germans as to their own behaviour, is enough to create the darkest presumptions as to how the trouble started, and would seem to place the burden of proof on them rather than on the Belgians.

While we were talking about this there came another rattle of fire, and we scuttled back to our shelter, among the horses. Every now and then a surly soldier with two huge revolvers came and looked over the ledge at us, and growled out: *Was machen Sie denn hier?* followed by some doubting remarks as to our right to be on the premises. As he was evidently very drunk and bad-tempered I was not at all sure that he would not decide on his own responsibility to take no chances and put us out of our misery. After several visits, however, he evidently found something else more interesting, and came back to trouble us no more.

When the row began a motor had been despatched toward Brussels to recall some troops that had left a few hours before. Now and then our officer came in to tell us what he thought of their chances of getting back.

On one of these visits, Blount remarked by way of airy persiflage, that that drink of wine that had been sent for was a long time coming. Anything as subtle as that was lost on our friend, for he walked solemnly away, only to reappear in a few minutes with a bottle and several glasses which he set up on the edge of the platform and filled with excellent Burgundy. We stood up among the horses and drained a bumper of the stuff, while the officer wandered back to his work. He had gone calmly out into the thick of things to rescue this bottle, and took it as a matter of course that we should claim the drink that had been promised us.

Presently, with a good deal of noise, a fairly large force of troops came marching down the boulevard, and took up positions around the station. Our officer returned, waving a smoking revolver, and told us to lie down as flat as we could among the horses, and not to move unless they got restive. He said it looked as though an attempt would be made to take the station by storm, and that there might be a brisk fight.

However, there were only a few scattering shots, and then our friend came back and told us that we had better get out and start for home before things began

A street of handsome homes at Louvain

Types of von Arnim's officers

A Landwehr infantry company

again. He added, however, that we must have the permission of the commanding officer who was on the other side of the station, but offered to pilot us to the great man and help us get the permission. The way lay straight out into the square, in full view of the houses across the way, along the front of the station just behind the troops and into the railroad yard on the other side.

That station seemed about four miles long, and the officer was possessed of a desire to loiter by the way, recounting anecdotes of his school days. He would walk along for a few steps and then pause to tell Bulle some long and rambling yarn about his uncle. Bulle would take him by the arm and get him in motion again. Then the old chap would transfer his conversational fire to another member of the party, and we were obliged almost to pull him the length of the square.

The commanding officer was a pleasant-faced little man who stood in the shelter of a water tank and received us in a puzzled way, as though he wondered what civilians were doing in that neighbourhood anyway. Permission was readily granted for us to leave, with the ludicrous proviso that we did so "at our own risk." Then Bulle put everybody in good humour by inquiring innocently if there was any danger. Everybody burst into peals of laughter, and we were escorted to our car by the same slow-moving officer, who insisted on exchanging cards with us and expressing the hope that we should meet again, which we could not honestly reciprocate. Then, after an hour and a half in the station, we got away amid a great waving of hands.

The boulevards were deserted save for the troops coming back into the town. New houses were burning that had been intact in the afternoon. After passing the Porte de Tirlemont, we began to see people again— little groups that had come out into the streets through a craving for company, and stood huddled together listening to the fighting in the lower part of the town.

In harmony with the policy of terrorising the population, the Germans have trained them to throw up their hands as soon as any one comes in sight, in order to prove that they are unarmed and defenseless. And the way they do it, the abject fear that is evident, shows that failure to comply with the rule is not lightly punished.

Our worst experience of this was when in coming around a corner we came upon a little girl of about seven, carrying a canary in a cage. As soon as she saw us, she threw up her hands and cried out something we did not understand. Thinking that she wanted to stop us with a warning of some sort, we put on the brakes and drew up beside her. Then she burst out crying with fear, and we saw that she was in terror of her life. We called out to reassure her, but she turned and ran like a hunted animal.

It was hard to see the fear of others—townspeople, peasants, priests, and feeble old nuns who dropped their bundles and threw up their hands, their eyes starting with fear. The whole thing was a nightmare.

We were dreadfully depressed, and rode along in silence until Bulle turned around from the front seat and inquired in a matter-of-fact voice:

"What sort of wine was that we drank at the station?"

We told him, and then he shook his head and said as though to himself:

"I drank a big glass of it, but I was so frightened that I didn't taste it at all."

That broke the edge of the strain we were under, and we had a good laugh in which Bulle joined.

And so back to the Legation without further mishap, to find everybody worrying about us, and the Minister putting his foot down and announcing that there were to be no more expeditions of the sort, no matter what the reason for them.

NOTE—The foregoing is an impression of one afternoon at Louvain, taken from a journal written at the time. It was intended to pass on the question of responsibility for precipitating the orgy of murder and bestiality indulged in by the German army from the 25th of August until the 30th, when orders were received from Berlin to stop the destruction and restore public order.

Many subsequent visits to Louvain, and conversations with people who were there when the trouble began, have only served to strengthen the impression that the whole affair was part of a cold-blooded and calculated plan to terrorise the civilian population.

While we were there, it was frankly stated that the town was being wiped out; that its destruction was being carried out under definite orders. When the German Government realised the horror and loathing with which the civilised world learned of the fate of Louvain, the orders were cancelled and the story sent out that the German forces had tried to prevent the destruction, had fought the fire, and by good fortune had been able to save the Hôtel de Ville. Never has a government lied more brazenly. When we arrived, the destruction of the town was being carried on in an orderly and systematic way that showed careful preparation. The only thing that saved the Hôtel de Ville was the fact that the German troops had not progressed that far with their work when the orders were countermanded from Berlin.

It was only when he learned how civilisation regarded his crimes, that the Emperor's heart began to bleed.

The true facts as to the destruction of Louvain will startle the world—hardened though it has become to surprise at German crimes. Unfortunately, however, it is impossible to publish the details at this time without endangering the lives of people still in Belgium under German domination. But these people will speak for themselves when the Germans have been driven from Belgian soil, and they are once more free to speak the truth.

During the afternoon Count Clary had come over and announced that Austria-Hungary had declared war on Belgium, and that he had to leave at once. He has turned his Legation over to us. I went around to see him late in the evening, and made the final arrangements. This afternoon the Danish Minister came in and turned his Legation over to us, as he expects to go in a day or two. That will make four Legations besides our own under our protection.

Austrian guns have been in action for some days, and now it has been thought worth while to regularise the situation. The Austrian Minister has, therefore,

Major von Herwarth (on the left) at the German outpost near
Hofstade

Monsieur Emile Francqui, President of the Executive Committee
of the Comité National de Secours et d'Alimentation

under instructions from his Government addressed the following note to the Belgian Government:

August 28, 1914.

"Whereas Belgium, having refused to accept the proposals made to her on several occasions by Germany, is affording her military assistance to France and Great Britain, both of which Powers have declared war upon Austria-Hungary, and whereas, as has just been proved (*no indication as to how or when it has been proved*), Austrian and Hungarian subjects in Belgium have been obliged to submit, under the very eyes of the Belgian authorities, to treatment contrary to the most primitive demands of humanity and inadmissible even toward subjects of an enemy State, therefore Austria is obliged to break off diplomatic relations and considers herself from this moment in a state of war with Belgium. I am leaving the country with the staff of the Legation, and am entrusting the protection of Austrian interests to the United States Minister in Belgium. The Austro-Hungarian Government is forwarding his passports to Count Errembault de Dudzeele.

CLARY."

This is the first we have heard of any mistreatment of Austrians in this country, but then they probably had to advance some sort of reason for going to war.*

---

* The Belgian Government sent the following reply to the Austrian declaration of war.

Antwerp, August 29, 1914.

Belgium has always entertained friendly relations with all her neighbours without distinction. She has scrupulously fulfilled the duties imposed upon her by her neutrality. If she has not been able to accept Germany's proposals, it is because those proposals contemplated the violation of her engagements toward Europe, engagements which form the conditions of the creation of the Belgian Kingdom. She has been unable to admit that a people, however weak they may be, should fail in their duty and sacrifice their honour by yielding to force. The Government have waited, not only until the ultimatum had expired, but also until Belgian territory had been violated by German troops, before appealing to France and Great Britain, guarantors of her neutrality, under the same terms as are Germany and Austria-Hungary, to coöperate in the name and in virtue of the treaties in defense of Belgian territory.

By repelling the invaders by force of arms, she has not even committed a hostile act as laid down by the provisions of Article 10 of The Hague Convention, respecting the Rights and Duties of Neutral Powers.

Germany has herself recognised that her attack constitutes a violation of international law, and being unable to justify it, she has pleaded her strategical interests.

Belgium formally denies the allegation that Austrian and Hungarian subjects have suffered treatment in Belgium contrary to the most primitive demands of humanity.

The Belgian Government, from the very beginning of hostilities, have issued the strictest orders for the protection of Austro-Hungarian persons and property.

DAVIGNON.

The —— Chargé came around this afternoon to ask about getting to Antwerp, where he wants to flee for protection. He was very indignant because the Military Governor had refused to allow him to go. When I asked him on what ground the permission had been refused, he said that it had not exactly been refused, but that he could go only on his own responsibility. He wanted us to protest against this. I meanly suggested to him that he would be in much more serious danger if he had an escort of German troops to take him to the Belgian lines, and he left in a terrible state of mind.

Mr. Whitlock and the Spanish Minister went to call on the Military Governor this afternoon to get off some telegrams which he had promised to send, and to talk over the general situation. After that they went to call on the Burgomaster, and came back with a pretty good idea of what was happening in our fair city.

The Governor loaded them up with a large budget of official news, showing that Germany was victorious all along every line; that she was not only chasing the French and English armies around in circles, but that Uhlans were within forty kilometers of Paris, and that five Russian army corps had been beaten in Eastern Prussia. It really looks as though things were going pretty badly for the Allies, but we have absolutely nothing but German news and cannot form an accurate opinion.

The Germans are particularly bitter against the Belgian clergy and insist that the priests have incited the people to attack the German troops and mistreat the wounded. So far as I can learn, this is utter rubbish. The authorities of the church have publicly exhorted the people to remain calm and to refrain from hostile acts, pointing out that any provocation would bring sure punishment from the German military authorities. The priests I have seen have been at great pains to set an example that the Germans should be the first to commend.

The clergy has a tremendous influence in Belgium, and is sincerely respected. They will be a vital force in holding the people together in their patriotic devotion, and in maintaining public tranquillity.

A new Governor-General is to be sent us here. The Minister could not remember his name. I am curious. von Lüttwitz will remain for the present at least.

The Burgomaster reports that the inhabitants of Brussels are calm and that there need be no fear of trouble unless they are allowed to go hungry. A committee has been formed to revictual the town, and is working night and day. Monsieur Solvay has given a million francs, and other Belgians given large sums. Soup kitchens are being started for the poor and the question of bringing food supplies from neutral countries is being taken up and pushed with activity. These Belgians are admirable in the way they handle things of this sort. They all realise the importance of keeping quiet so as to avoid any possibility of a repetition of the Louvain business. It would take very little to start something of the sort here and the result would be the same—the destruction of the city. Naturally everybody is turning to and trying to head off any excuse for violence.

Brussels, Sunday, August 30, 1914.—Our place has got to be the local diplomatic corner grocery, where all the village loafers come to do their heavy loafing. They bring in all the fantastic rumours that are abroad in the land, and discuss them with all solemnity. In the last day or so we have had it "on the best authority" that the Queen of Holland has had her consort shot because of his pro-German sympathies; that the Kaiser has given up all hope and taken refuge in Switzerland; that the United States had declared war on Germany and Austria; that the King and Queen of the Belgians had fled to Holland, and that all was over. These are just a few.

Troops have been pouring through the town steadily all day on the way to Vilvorde, where the Belgians are still fighting, and to the south, where there is heavy cannonading going on. The Belgians are making a big fight on the Antwerp road, evidently to hold the attention of one German army corps and lighten France's load by just that much. It is a hopeless fight so far as they themselves are concerned, but it throws their courage and fighting qualities into higher relief.

We are now the pampered pets of both sides. The German troops cheer our flag when the motor noses its way through them. The people of the town are equally enthusiastic, and many of them are wearing small American flags in their buttonholes. How long it will last there is no telling, but while it does, our work is made just that much easier.

Lunched at the Palace Hôtel with Bulle and Blount. Riseis, the Italian Secretary, came in and joined us. Bulle told him the story of our trip to Louvain with embellishments that made my eyes start from my head. I had not realised what a desperate adventure we had been on until I heard it as it should be told. It made the real thing seem mild.

Before lunch we drove to Blount's to learn whether the cannonading to the south was still going on. It was—heavy booming of German guns; no French guns to be heard. Late in the afternoon Blount and I drove off into the country to see whether we could locate the fighting to the south. We got as far as Nivelles, but all was as peaceful as it should be on a perfect Sunday afternoon. The people there were surprised that anyone should have thought there was fighting there. It was still much farther to the south. We drove around in search of evidence of fighting, but could find none. And this after circumstantial accounts of hand-to-hand struggle through all this part of the country!

August 31st.—This morning began with a troop of people in to tell us that the rough work was about to begin, and that Brussels was to go up in smoke. There is a good deal of unrest in the lower end of town and trouble may break out at any time. Bad feeling has grown a good deal in the past few days and one good row would throw the fat in the fire. I went through the rough part of town late this afternoon and found patrols everywhere, heavily armed and swaggering about in groups of four. For their own sake I hope the people will not do anything foolish.

People are making another effort to get away and are not finding it easy. At six this morning a crowd left here for Ninove, twenty kilometers to the west. Twenty-five hundred of them clung all over the trams that make the trip. At Ninove they walked a mile or so, carrying their belongings, and caught a train to Alost, where they changed for another train for Ghent. Goodness knows how many changes they had ahead of them after that. The trip was supposed to end safely in Ostend some time this evening. It usually takes two hours.

Hearing that the train service was open and that boats were running from Ostend to Folkstone, we decided to verify the tidings and then get off some of our people, who should have gone long ago.

To make sure Blount and I motored down to Ninove after lunch to telephone the Consul at Ostend and learn the true state of affairs. When we reached Ninove we found the station so packed with refugees that there was no getting near the telephone bureau. The Chef de Gare, who had never in his long and honourable career had such a mob to lord it over, was so puffed up that he could not get down near enough to earth to hear our questions, so we decided to proceed to Alost and try our luck there.

We motored over in short order and got quick communication with the Consul at Ostend. He had very little news save that a lot of British Marines had been landed there and had to-day been taken away again. He gave us what we wanted in the way of steamer information.

I got the Consul-General at Antwerp on the telephone and learned that all was well there.

As I came out of the booth from this second call, I was held up by a Garde Civique, who inquired if I was

the *Monsieur de l'automobile*. He would like to see
my papers. Certainly. Then I remembered that I
had left all my Belgian papers at the Legation and had
nothing but papers in German from the military
authorities. I showed them anyway. Before he could
examine us any further, three eager amateur Sherlocks
came bursting into the room and took charge of the
proceedings. The leader pointed an accusing finger at
Blount, and exclaimed, "You have come from Ninove!"
Blount admitted it. "You had a third person in the
car when you left there!" "*Pas du tout.*" "On the
contrary, I have three witnesses to prove it." Aside
from the fact that nobody could have got to Alost in
the time we had, it made no real difference how many
people we had in the car, and Blount said as much.
Then our accuser changed his plan of attack. "I
observed you when you arrived, and you were speaking
a language which was perhaps not German, but sounded
like English." "It was," said Blount. "Aha," trium-
phantly, "but you said you were Americans!"

By this time the Chef de Gare had come to answer
our questions and we waved our persecutors aside
while we talked to him. They kept quiet and meekly
stood aside, as we bade them. While we talked with
our functionary, I looked out on the square and saw
that we were a real sensation. The Garde Civique
had been called out and was keeping the place clear.
The crowd was banked up solid around the other three
sides of the square. They looked hopeful of seeing
the German spies brought out and shot. By signing
our names on a scrap of paper, which the amateurs
compared with the signatures on different papers we

had about us, we convinced them that we were harmless citizens, and were allowed to go. The crowd seemed greatly disappointed to see us walk out free. The Garde Civique let them loose as we got in the car, and they came thronging around for a good close look at us.

We honk-honked our way through them, thanking our lucky stars we had not had a worse time of it.

At the edge of the town we looked up and saw two German aeroplanes snooping around. A minute later a crowd of people surged across the street to bar our way, shouting that we must go no farther, as the Germans were approaching the town and that it was dangerous to proceed. Two young officers came across the street to tell us in great glee that they had made a dash in a motor at the first German outpost and had brought in four prisoners. They were bursting with joy in their exploit, but by this time they may themselves be prisoners.

In a few minutes we came to the first German outpost, and had our papers carefully examined. From then on we were held up every few yards and nearly had our papers worn out from much handling. At one place a young Lieutenant looked over our papers and burst out into roars of laughter at the name of von Jarotzky. He called to other officers. They came up, looked at the signature, and also burst out into loud laughter. I asked them what the joke was, but they were not telling.

We got in about seven o'clock, without incident. Went to see von Herwarth after dinner on behalf of a poor Belgian woman whose husband, a Major in the Grenadiers, is dangerously wounded and in the military hospital at Antwerp. The Germans are going to send her up to-morrow on a motor with some Belgian officers, who are being exchanged. I saw the aide-de-camp who is going through with the car and asked him to be nice to her. Then to her house, to shut up a lot of old women of both sexes who were trying to dissuade her from going, on the ground that the Germans would hold her as a hostage. I suppose she will be off.

Mrs. Bridges,* wife of the former British Military Attaché, was in this evening for help. A British prisoner told of seeing Colonel Bridges fall from his horse at Mons, mount again, ride a little way and fall. She cannot get to Mons, so we are getting her off to France via England, in the hope that she may find him on that side.

It is a pitiful business, and the worst of it is that they all think we have some miraculous power to do anything we like for them. I only wish we could.

Brussels, September 1, 1914.—The first thing this morning I had a pow-wow with Hulse about how to handle the funds that are being gathered to relieve the enormous amount of distress that we shall have to meet here. There is a good deal of it even now. All the big factories are closed. Most of the shops have their shutters up, and the streets are filled with idle people. Importations of foodstuffs, even from the outlying districts, have stopped dead. Conditions are bad enough

---

* Colonel Bridges was badly wounded at Mons, but escaped, recovered, was wounded again at Nieuport, but survived both, and having received the rank of Lieutenant-General, was the military member of the Balfour Mission to the United States in 1917.

in all conscience, but they are nothing compared to what we have ahead, when cold weather comes on.

A lot of bankers and big business men have got together to wrestle with the financial problem. The Burgomaster has his people at work, trying to get their hands on foodstuffs and coördinate their work.

I went to the Foreign Office and talked things over with von Herwarth. He straightened out some of the tangles, and we were able to get things moving.

I have no trouble with the higher officers at headquarters, but I never go there that I do not want to murder the large brutes of non-commissioned officers who guard the door. They wear large brass plates on their chest and look like bock-beer signs. They have a free and careless way of booting everybody out the door and refusing to listen to anybody. I get fighting mad every time I go there and this morning got sufficiently roused to develop considerable fluency in German. I pictured to the large rough-neck some of the things that were going to happen to him if I was not let in; he was sufficiently impressed to permit me to stand on the sidewalk while my card was sent in. When I got in I made a few well-chosen remarks on the manners, if any, of the watch dogs of the Ministry.

From the Ministry I went to the Société Générale, where I was asked to attend a conference between the bankers of the city. There were ten of them in the big directors' room, and they worked to some purpose. M. Francqui, the director and leading spirit of the Société Générale, presided over the meeting. He explained the general situation simply and clearly, and stated what they had done and wanted to do. They had three points on which they wanted advice, and they were brought up and disposed of one at a time. By twelve o'clock I got away, and felt that the hour I had put in there had been well spent.

When I got back to the Legation, I found a nice Belgian who had no request to make of us, but wanted to tell his story to somebody, and a terrible story it was, too. He had fitted up his château near Mons as a Red Cross hospital. During the battle there a week ago, 102 British wounded had been brought in. The Germans found the château a hindrance in their operations, so got it out of the way by battering down the walls with artillery, and then throwing grenades into the building to set it on fire. There was great difficulty in getting the wounded out and hiding them in such shelter as was to be found. One man, at least, was burned alive in his bed. It seems incredible that Red Cross hospitals should be attacked, but stories come in from every side, tending to show that they are.

Beside this man's property there is a railway crossing. When a troop train passed over it day before yesterday, there was an explosion like the report of a rifle. The train was immediately stopped. The officer in command announced that civilians had fired upon his train, and ordered all the men in the vicinity taken prisoners. Then, refusing to listen to explanation or discussion, he had them all stood up against a wall and shot. When it was all over, he listened to explanations and learned that the report was that of a cap placed in the switch by the German railway men as a signal to stop the train before reaching the next station. By way of reparation, he then graciously admitted that the civilians were innocent. But, as my caller said: "The civilians were also dead."

Another pleasant thing the Germans seem to be doing is arresting peaceful citizens by hundreds and sending them back to Germany to harvest the crops. They will also reap a fine harvest of hatred for generations to come.

Poor Bulle is in considerable doubt as to his status. For many months he has not heard from his Government, if any, and has not been able to get a word as to whether he is Chargé d'Affaires or not. I told him to-day that he had a rather unique situation as the representative of a country without a Government to a Government without a country. He extracted a chuckle from that.

Blount made up his mind to leave for America this afternoon, by way of Ostend and England. His family was all ready to start, but when he went down to headquarters to get a *laisser-passer* it was refused. Operations are apparently about to be started in *tout le bazar*, and they don't want stray civilians seeing too much. Blount will now settle down here for the present. His loss is our gain.

The Danish Minister was in again this afternoon. He is going away, and has finally turned his Legation over to us. We now have four Legations besides our own—German, British, Austro-Hungarian, and Danish.

One little thing the Germans have done here that is *echt Deutsch* is to change the clocks on the railway stations and public buildings to German time. Every other clock in town continues about its business in the same old way, and the change only serves to arouse resentment.

Another thing is, that on entering a town, they hold the Burgomaster, the Procureur du Roi and other authorities as hostages, to ensure good behaviour by the population. Of course the hoodlum class would like nothing better than to see their natural enemies, the defenders of law and order, ignominiously shot, and they do not restrain themselves a bit on account of the hostages. Just lack of imagination.

September 2nd.—A paper, smuggled through the lines from Antwerp this morning, gives the news that the Queen has left for England, with the royal children; adding, "she is expected back in a few days." This move is evidently in anticipation of the bombarding of Antwerp.

Now and then a Belgian has the satisfaction of getting in a gentle dig at the Germans; although, if the dig is too gentle, the chances are the digee does not know it. Last week Countess Z——, aged eighty-four, who is living alone in her château, was obliged to put up a German General and his staff. She withdrew to her own rooms, and did not put in an appearance during the two or three days that they were there. When the time came for them to leave, the General sent word that he would like to see her. She sent back a message, asking to be excused. The General was insistent, however, and finally the little old lady came reluctantly down the stairs into the great hall, stopping three or four steps from the bottom and gazing down upon her lodgers with a quizzical smile. They all clicked their heels and bowed, and then the General stepped forward a few paces and, in his best manner, said that they could not go away without thanking her for all that had been done to make them comfortable during the time they had had the honour of being her guests. When he had quite finished, the little old lady replied in her gentle soft voice:

"*Messieurs, vous n'avez pas à me remercier. Je ne vous avais pas invités.*"

Brussels, September 2, 1914.—A beautiful aide-de-camp of Field-Marshal von der Goltz turned up this afternoon, and announced that, if agreeable, His Excellency the Governor-General, would call to-morrow afternoon between four and five. We are looking forward with a good deal of interest to seeing the big man. He arrived yesterday, but has kept so quiet that nobody knew he was here. The aide-de-camp nearly wept on my shoulder; said there was nobody in the General's party who knew Brussels, and that they were having a terrible time to find their way around the town. He'll probably have greater worries before he gets through.

We have at last heard from McCutcheon, Cobb, Lewis, Bennett, etc. A telegram came to-day from the Consul at Aix-la-Chapelle, asking that we look after their baggage at the Palace Hotel. From this we judge that they were arrested and sent back to Germany on a troop train. They left here for Mons, and goodness only knows what adventures they have been through since we last saw them.

Brussels, September 3, 1914.—This afternoon, at four o'clock, von der Goltz (Field-Marshal Baron von der Goltz Pacha, to be exact) arrived with a staff of seven officers to make a formal call. A crowd quickly gathered in the street, as their big gray military cars snorted up to the door. All the neighbourhood was in a great state of excitement. The great man is pretty old and doddery, wears spectacles about an inch thick, and a large collection of decorations. His staff was also brilliant in decorations and silver helmets, etc. I met them at the foot of the stairs, and escorted them up. The Marshal is apparently blind as a bat, for he never turned on the landings and would have walked straight into the walls if I had not steered him around the corners.

After one good look we decided that he was to be a figure head and leave the real work to the troop of officers and functionaries he had brought with him.

It was supposed to be a purely formal call, but the old gentleman seemed to have no thought of leaving, and did not budge for half an hour. The conversation was not thrilling.

They finally left after much clicking of heels, and the bemonocled Count Ortenburg nearly broke his neck by tripping over his sword. However, we got them safely out of the house, while all the servants leaned out of the windows and took in the show.

The new Governor - General has addressed a Proclamation to the Belgian people, and has had it posted on the walls:

PROCLAMATION.

His Majesty, the Emperor of Germany, after the occupation of the greater part of Belgian territory, has been pleased to appoint me Governor-General in Belgium. I have established the seat of the General Government in Brussels.

By His Majesty's orders, a civil administration has been established with the General Government. His Excellency Herr von Sandt has been made Chief of this Administration.

The German armies advance victoriously in France. My task will be to preserve quiet and public order in Belgium.

Every act of the population against the German military forces, every attempt to interfere with their communications with Germany, to trouble or cut railway, telegraph or telephone communications, will be punished severely. Any resistance or revolt against the German administration will be suppressed without pity.

It is inevitable in war that the punishment of hostile acts falls not only upon the guilty but also on the innocent. It is the duty of all reasonable citizens to exercise their influence with the turbulent elements of the population to restrain them from any infraction of public order. Belgian citizens desiring to return peaceably to their occupations have nothing to fear from the German authorities or troops. So far as is possible, commerce should be resumed, factories should begin to work, and the crops harvested.

BELGIAN CITIZENS

I do not ask any one to forego his patriotic sentiments, but I do expect from all of you a sensible submission and absolute obedience to the orders of the General Government. I call upon you to show confidence in that Government, and accord it your co-operation. I address this summons particularly to the functionaries of the State and of the communes who have remained at their posts. The greater your response to this appeal, the greater the service you will render to your country.

The Governor-General,

Brussels, September 2, 1914.   BARON VON DER GOLTZ,

Field-Marshal.

Field-Marshal Baron von der Goltz Pacha, First Governor-General
in Belgium

The signatures to the "scrap of paper"

At about five o'clock, Bulle came along, and we went for a long walk together—the first time I have tried anything of the sort since the war began. We tramped out to the Bois and made a swing around the circle, not getting back until half-past seven, when we repaired to the Palace Hotel and had dinner with several of the colleagues. When von der Goltz left us, he had started for the Spanish Legation; but we learned from the Spanish Secretary that he had never arrived. Instead, at the last minute, an aide-de-camp had come clanking in to express His Excellency's regrets that he was unable to come, and say that he would have to defer his visit until a later date. Something happened to him after he left our Legation.

X—— had an experience yesterday which made him boiling mad. He left town in the afternoon with his Consul, to go to Alost for telegrams and letters. He was in a car flying his flag, and had his *laisser-passer* from the German military authorities. Near Assche, he was stopped by an outpost, and told he could not go any further. He accepted this in good part, and said he would go back. At this point, an old turkey gobbler of a General arrived and lit into him for being there. He replied that he had done nothing to which exception could be taken; that his papers were in order, and that he was ready to return at the first indication from the military authorities. This seemed to enrage the old soldier who announced that they would do nothing of the sort; that they were prisoners of war and would be sent back under armed guard. X—— protested that this was an outrage against the representative of a friendly country, but in spite of this two armed soldiers were placed in the car with them and another beside the driver, and they were brought back to town as prisoners. By dint of arguments and threats they were taken to headquarters instead of jail, and succeeded in seeing General von Lüttwitz who piled on the excuses. It does you no good to have legitimate business and papers in order if it suits some apoplectic officer to clap you into jail.

One of the officers I saw to-day told me that the Germans were deliberately terrorizing the country through which they passed. It is a perfectly convincing explanation of German doings in this country, but I did not think they were prepared to admit it so frankly. This frank fellow made no claim that civilians had attacked the German troops; his only observation was that they might do so unless they were so completely cowed that they dared not raise their hands. He emphasised the fact that it was not done as a result of bad temper, but as part of the scheme of things in general. For my information, he remarked that in the long run this was the most humane manner of conducting war, as it discouraged people from doing things that would bring terrible punishment upon them. And yet some of these Belgians are ungrateful enough to complain at being murdered and robbed.

September 4th.—Autumn is coming with little gusts of wind and falling leaves. Clouds are thick, and there is a sort of hidden chill in the air. It is depressing in itself, and makes us think with some dismay of what is ahead of the millions of men who are in the field, if the war is to continue into the winter as seems probable.

I am sure there is something big in the air to-day. For several days there has been a growing nervousness at headquarters. For four days there has been no official proclamation of German victories. Persistent rumours come in of large numbers of British troops between here and the coast, advancing in the general direction of Brussels. X——'s arrest, while on a trip to Alost, looks as though the Germans had some reason for keeping people from getting out that way with knowledge of military conditions here. Another thing. We were to have returned the call of von der Goltz to-day at noon. Between here and the Spanish Legation yesterday, *something* happened. He never got to the Spanish Legation. This morning we got a message from the Etat-Major that von der Goltz had "telegraphed" to ask that we should postpone our call. Where he is, nobody would say. The officer who brought the message merely stated that he had been called away in great haste, and that it was not known when he would return. Troops are marching through the town in every direction, and in large numbers. Supply trains and artillery are creaking through the place night and day, and we are awakened nearly every morning either by the crunching of the heavy siege pieces or the singing of large bodies of troops as they march through the streets. Every day we realise more and more the enormous scale on which the operations are being conducted. It seems tremendous here, and we are seeing only a small part of one section of the field of operations.

Privately, the Germans continue to assure us that they are winning all along the line. They say that they have taken the whole of the first line of defences in France with the single exception of Maubeuge, where there has been long and heavy fighting and where the result still trembles in the balance. In addition to this they claim to have taken a part of the second line of defences. They say that the French Government has removed to Bordeaux, which seems quite possible, and even sensible. They tell us all these things every time that we go over to the General Staff, but they do not publish anything about it.

A British Red Cross doctor was in to-day and told us some items of interest. He said that he had been assigned to care for the wounded prisoners who were being brought back from France on their way to Germany, and that he had seen all the British prisoners who had been brought back by way of Brussels—about three thousand in all. He said that they were in good spirits and were sure that things would come out right in the end. There were the remnants of the Argyle and Sutherland Highlanders, who went into action something over a thousand strong and came out only a handful.

I made two attempts to see Herwarth to-day but was kept on the sidewalk and in the courtyard by the big green dragons who guard the entrance to head-quarters. After the second attempt I returned to the Legation and telephoned him that I should like to see him when he could get it through the heads of these people that we were not tramps. He was very nice and apologetic and had all the officers in the German army out in the street waiting for me when I went back for the third time. All the sentries were blown up and given the strictest sort of instructions that I was to be passed along without question whenever I appeared. I was also given another *Passierschein* to add to my collection, directing everybody to let me pass wherever I wanted to go. In view of the fact that a lot of our work here is in behalf of German subjects, this is about the least they could do.

Some news has been brought down from Antwerp that makes it seem necessary for me to go there and get back again before the siege begins. I had hoped to get away this morning but have not yet been able to get a decision as to exactly what is to be done. I now hope to get away after lunch.

I spent all yesterday afternoon enciphering a telegram which I must get off either through Holland or Antwerp. We are able to send nothing but open messages over the military wire through Berlin and I have a strong suspicion that these are being censored.

Brussels, September 7, 1914.—Did not get off to Antwerp to-day but hope to make it by to-morrow noon. There was too much going on, but arrangements are being made for a *laisser-passer*, etc., and I anticipate no trouble beyond being shot or made prisoner.

Gherardi* came in this morning for a call and then left for Maubeuge, which the Germans had arranged to capture during the day. They seemed very sure of it, but I would not be surprised to see him come sailing back without having seen the surrender.

Baron von der Lancken, of the Foreign Office in Berlin, called this morning. He is here to handle

---

* American Naval Attaché at Berlin.

relations with the Foreign Ministers remaining in Brussels. As we have had the care of German interests they all come here first and our position is better than that of any other Legation in the country. We have things on a working basis.

September 8th.—Last night, after dinner, I trotted around and called on the wives of some of the Belgian officials to see whether there was any news of them that I could give to their husbands in Antwerp. I found Madame Davignon, the wife of the Minister for Foreign Affairs, in her son's home, peacefully working away on clothes for the wounded. She told me all the news of the house so that I could repeat it to her husband. She is as calm as you please and far from despairing.

Madame de Broqueville, the wife of the Prime Minister, turned her house into a Red Cross hospital at the outbreak of hostilities; it is a beautiful big place. Of course there are practically nothing but German wounded in the house now, but the good lady conquers her natural feelings and has them as well looked after as though they were of her own race. I went in in an apologetic mood for intruding on her at so late an hour, but she had lots to say and I stayed on for a long time. It did her good to talk, and I was so overawed by her courage and poise that I sat and listened in silent admiration. The wives of the Cabinet Ministers and other officials have shown wonderful nerve and are standing right up to their duty.

Count and Countess de X had an interesting story to tell of their experiences when the first armies went through. When the war broke out they were at their château and were caught by the first onrush of troops. Their fine cellars were emptied for the benefit of the invader, but nothing more serious happened to them until the second wave came along. Then there was a demand for more wine. As all the wine had been carried away they could not comply. The Germans were convinced that they were being fooled, and searched the place very carefully. Finally they imprisoned the X's for three days in the cellar and then brought them forth and stood them up before a firing squad and threatened to shoot them unless they told where the wine was hidden. At the critical moment a big gray military car rolled up, and to their considerable relief they saw that one of the occupants was a German princeling, who had formerly been their guest on several occasions. They called out to him, and by his orders were immediately released. After expressing their thanks to him they went into the château to find that soldiers were engaged in packing up their fine collections of enamels and porcelains to ship them to Germany. Another appeal to the Prince, who was most sympathetic. He was a practical and resourceful man, and said:

"Of course I'll stop this, but you will understand that our men would like to keep some little souvenir of the war in Belgium. That would be hard to prevent. But I would suggest that you pick out all the pieces that you value most and pack them away in that large wardrobe. Then I'll do the rest."

Madame de X was, of course, delighted with this, and scurried about gathering together the finest pieces and packing them carefully into the big wardrobe. She kept it up as long as there was a nook or cranny where odd pieces could be put, and then reported progress to the Prince.

"Are you sure that all the best pieces are there?" says he.

"All that could be packed there," answers Madame de X.

"Good," says the Prince, and then turning to his orderly: "Have that wardrobe sent to Berlin for me."

The way the German army cleaned out the wine of the country was a revelation to everybody. They would not take what they needed for the day's drinking but would clear out whole cellars at a time and load what was not drunk onto carts to be carried away. The result was that people who had a little warning had recourse to all sorts of ingenious tricks to save some of their store. There was one bright man in the province of Namur who removed his stock of wine— all except a few thousand bottles of new wine—and deposited them in the ornamental pond near his château. The Germans arrived a few hours afterward and raised a great fog because they were not satisfied with the amount of wine they found. The owner of the château had discreetly slipped away to Brussels and they could not do anything to him. However, they tapped all the walls for secret hiding places and went over the park to see if anything had been buried—all in vain. The next morning, however, the pond was covered with labels which had soaked off and floated to the surface, and after draining the pond the whole stock was carted away.

Madame B——, who was there, has an interesting souvenir which she proposes to keep if possible. Dur-

ing the first days of the war her château was occupied
by a lot of officers, who got gloriously drunk and
smashed up pretty well everything in the drawing-
room and dining-room. One of them, with a fine sense
of humour, took a piece of hard chalk and wrote on
the top of her piano in large letters: *Deutschland über
alles!* The crowd left the place in the morning without trying to cover their traces, and Madame B——
came in to put things to rights. The first thing she
did was to get a large piece of plate glass to cover the
top of the piano so that the legend would not be effaced,
and over that she placed an ordinary piano cover so
that no future visitor would be inclined to erase the
inscription. When the war is over this will be an
interesting reminder of her visitors.

This morning I was ready to start for Antwerp.
My *laisser-passer* had been promised for ten o'clock.
When it did not come by that hour, I went up to see
Baron von der Lancken who had agreed to attend to
the matter. He received me most graciously, told me
how delighted he was to see me, how it pleased him
to see that we came to him with our little troubles, etc.
He kept off the subject of the *laisser-passer* as long as
he could, but when he could stave it off no longer he
said that he must ask me to see von Herwarth, who had
been placed in charge of all matters regarding passports, etc. I made a blue streak over to Herwarth's
office, and saw him after a little delay. He kept me
as long as he could, and told me all that he knew about
the war and perhaps a great deal more. When we got
down to the subject of my visit he said that von der
Lancken was mistaken, that passports could be granted

only by Colonel von Claer who had his office about a block away. I began to smell a rat about this time, but kept plugging away. I spent an hour and a quarter in the antechamber of the Colonel, being unable to get to him or to any of his officers. It was all part of a game. Both von der Lancken and Herwarth harped upon the danger of the trip to Antwerp, advised against it and told how terribly they would feel if anything were to happen to me. I asked each of them point blank if they contemplated an attack while I was there. They both avoided the subject, but said that with the situation as it was now it was impossible to tell from one moment to another what might happen. I saw that they were undecided about what was going to happen next, and that until they did know they did not intend to let me go. They naturally do not wish to have anything happen to me or anyone else connected with the Legation, so I feel entirely safe about going.

After lunch I went back to the siege and stayed until my friend, the Colonel, left by the fire-escape or some equally desperate way so as to avoid seeing me.

Von der Goltz had sent word to the Minister that he was coming here for tea this afternoon, and wanted to meet the Spanish Minister. That was our opportunity, and the Minister was all primed with what he was to say to the old chap. They beat us to it, however. The problem had evidently been decided since I saw von der Lancken in the morning, for he greeted me with the news that the *laisser-passer* would be around in the course of the evening. He added that the General was anxious to send one of the Belgian

Ministers of State to Antwerp, and would appreciate it if I would take him with me. He is Count de Woeste, the man who has always fought against having an army, on the ground that Belgium was so fully guaranteed by her treaties that it was unnecessary. Baron von der Lancken says that they will make out a *laisser-passer* on which he will be included, and that the military authorities will mark out the route by which we had best go, so as to avoid running into trouble. I imagine it will take us by way_of Termonde and St. Nicolas.

The crowd that came to tea included von der Goltz, Pacha, Baron von der Lancken, Herr von Sandt, and Count Ortenburg—a scion of a mediatised Bavarian family. They told us of all the glorious triumphs of the German army, and of the terrible drubbing that was in store for their enemies. They stayed on for about an hour.

When they left, I escorted the old man to his car. Before he climbed in, he looked me over curiously and remarked: "*Tiens, c'est fous qui faîtes ce foyage à Anfers! Four afez peaucoup de gourage. Che tacherai d'arranger un petit entr'acte pour fous être agreaple. Mais il vaut refenir aussitot gue bossible!*" They evidently intend to hold off for a day to await certain developments, and I am to get the benefit of the delay.

The Marshal also told us that Maubeuge had fallen, and that they had made forty-five thousand prisoners. It seems almost incredible that the French and English would have left that many men at Maubeuge when they knew that it was bound to fall.

Perhaps we shall find that this is not altogether accurate. They say nothing about what is happening in Austria. The news from England and Antwerp is to the effect that the Russians are giving the Austrians a hard time of it.

This afternoon the German headquarters issued an order prohibiting the bringing of newspapers to Brussels from the outside world, and announcing that any one who brings newspapers here or is found with papers in his possession will be severely punished. Two German papers will be distributed by the authorities, and everything else is taboo. They evidently intend that their own version of passing events shall be the only one to get out here.

Brussels, September 13, 1914.—Ever since the 9th I have been off on my little jaunt to Antwerp, and have not been able to get a line on paper.

I was not at all sure that I was going to get away at all, until I got down to the Legation on Wednesday morning and found my *laisser-passer*, signed by von der Goltz, waiting for me—another to add to my already large and interesting collection. With it was a letter from my friend and well-wisher, Baron von der Lancken, who said that an officer would be assigned to accompany us as far as the German outposts. He suggested that I take along a large white flag to be hoisted over the motor for the run between the lines. The note and *laisser-passer* had arrived at the Legation about one o'clock in the morning, and had looked so important that the slaves waked the Minister from a deep sleep to receive them.

Passierschein

für den ersten Sekretär der Gesandtschaft der Vereinigten Staaten
von Amerika zu Brüssel

    herrn Hugh G i b s o n

und den Königlich Belgischen Staatsminister

    herrn W o e s t e ,

die sich nach Antwerpen begeben, von wo sie am 10.September d.J.
nach Brüssel zurückzukehren beabsichtigen.

    Bei der Rückreise von Antwerpen nach Brüssel werden
die vorbezeichneten Herren begleitet sein von dem ersten Sekretär
der hiesigen Spanischen Gesandtschaft,

    Marquis de F a u r a

und einem oder zwei weiteren belgischen Herren, deren Namen noch
nicht angegeben werden können.

    Die Herren sind frei und ungehindert passieren und
repassieren zu lassen. Jeder Beistand ist ihnen zu gewähren.

            Brüssel, den 9.September 1914.

        Der Generalgouverneur in Belgien.

                    Generalfeldmarschall.

Pass issued by Field-Marshal von der Goltz to enable
Mr. Gibson to pass through the German lines to
Antwerp.

When I got to the office I found that Villalobar had not sent over his contribution of letters, so I ran up to the Legation and saw him. He bade me farewell as though I were off to certain death, and loaded me with a large bundle of letters and telegrams.

When I got back to the shop, I found my fellow-passenger, the Count de Woeste, waiting for me. He is a leader of the Catholic party which has been in power in Belgium for the past thirty years, and, although he is seventy-five years old, he is still a big figure in the little country. He behaved very well on the trip, and if I were a Belgian citizen I should vote for him on account of his good nerve.

We bowled off to headquarters, where I was mightily pleased to find that von Herwarth had assigned himself to the duty of taking us up to the outposts—just for a visit. It was the only satisfactory one I have had with him since he came. At headquarters there were always too many interruptions. My old travelling companion had a hard time to keep himself in hand and not enter upon a joint debate upon the war, its causes and justification. He did well, however, and my two passengers parted on good terms, even going to the extraordinary length of shaking hands at the outpost.

A big military motor, filled with armed men, was sent ahead to act as guide, and we followed along closely behind in a cloud of dust.

From the outskirts of Brussels right up to the German outposts at Hofstade, the fields were filled with German troops of every sort—infantry, lancers, heavy artillery, and even three or four large detach-

ments of sailors in blue blouses and caps. All the
men, except the sailors and a few of the Landsturm
who wear conspicuous blue uniforms, were in the new
greenish grey, which is about the finest color that
has yet seen active service. Frequently we drove
several hundred yards beside a field before noticing
that it was filled with soldiers. Several of the villages
between Dieghem and Hofstade were partially burned,
and there were evidences of shell fire—which to these
peasants must be a perfectly convincing substitute for
hell-fire—and of fighting at really close quarters.
Between Perck and Hofstade, the fields were covered
with deep entrenchments, and over some of these were
stuck dummy heads to draw hostile fire. Some, on
the other hand, were fitted with Belgian caps picked
up on the battle-field, evidently for the purpose of
inducing Belgian troops to approach for a closer look
before firing. Most of the big trees along the road
had been cut down, and many houses razed to the
ground so as to have a cleaner sweep for the artillery.
At Dieghem, the German pilot-car picked up a naval
officer who was to accompany us as far as the outposts
and to inspect his men on the way back.

On the outskirts of Hofstade, under a brick railway
bridge, we found the last German troops. They had
some hard fighting here at the time of the last Belgian
sortie, and the bridge and the surrounding houses
showed evidences of shell fire.

I was rather against putting up the white flag, but
both Herwarth and the naval officer were most insistent
that I should do so, saying that the country between
the lines was filled with patrols, both Belgian and

A street in Louvain

Fixing on the white flag for the dash between the lines

Refugees from the villages near the Antwerp forts

Arrival in Antwerp of refugees from Malines

German; that they felt that hostilities were to be commenced at any moment, and that any one who ventured into the district between the lines would stand a fine chance of being shot unless he carried a conciliatory emblem. They rigged up a long pole on the side of the car with a white flag about six feet square, and bidding a glad farewell to the representatives of Hohenzollern and Company, we started out to feel our way into Malines. About 500 yards beyond the bridge we sighted two Belgian bicycle patrols who, on seeing us, jumped off their machines and ran into an abandoned farmhouse. Knowing that they were at high tension, we crept up very slowly so that they might have a good look at us before trying their marksmanship. They were peeking over the window-ledge, with their rifles trained at us; but after a good look at the black clothes and white whiskers of M. de Woeste they pulled in their weapons and waved us to go ahead. About a kilometer farther on, we came around a turn in the road and nearly ran into the first Belgian outpost—six men and an officer. As we came around upon them they scurried behind stone walls and trees, and gave us the usual pleasant greeting of levelled rifles. As the most prudent things to do under such circumstances, the car was stopped, and I went ahead to parley. The officer proved to be young Z——. He turned quite white when he got a good look at me, and remarked that it was fortunate they had not had a sight of us farther down the road, as we would certainly have been filled with lead.

He said that the Germans had tried three times that morning to get through the lines in cars flying the white flag, in one instance at least, with a machine-gun in the car. As a result of this, the outposts had orders not to take any chance for the rest of the time intervening before the attack which was expected to begin at any minute.

Far be it from me to suggest that our friends had me put up the white flag, so as to offer proof of the Belgian savagery in firing on the white flag.

After this little experience, we took in our white flag and made the rest of our trip without trouble. We found outposts about every hundred yards, and were stopped at the point of the rifle each time; but as we got farther away from the outer lines the behaviour of the posts was noticeably less nervous, and when we got into Malines the mere sight of our papers was sufficient to let us freely through.

Since my last trip, the Belgians have been working steadily at their preparations for defence, and have accomplished wonders. Their large tracts of land, some of them forming natural routes, for entry between the forts, have been inundated with water from the canals so as to be quite impassable. Tremendous barbed wire entanglements form a broad barrier all around the outer and inner fortifications; they are so thick and so strongly braced that artillery fire would be practically useless against them, and cutting with wire nippers would be so slow that it could not be accomplished without a horrible loss of men.

There are any number of huge searchlights placed on the fortifications to sweep the skies for Zeppelins. Since my last visit, one Zeppelin had succeeded in getting over the town, but was surprised and dropped its whole cargo of 15 bombs in a distance of a few hundred yards, taking no lives and doing little material damage. Since then, several big craft have appeared at night, but have always been frightened away by the searchlights and the fire of the small vertical guns which have been ready for them.

All the villages which cluster around the fortifications have been razed to the ground, and the avenues of big trees have been cut down; it is a pretty dreadful sight.

I left M. de Woeste at the Grand Hôtel, where the Cabinet is staying, and then made for the Saint Antoine. Had lunch with Sir Francis Villiers and Colonel Fairholme, and got my first real news since the Prussian headquarters stopped issuing bulletins of German victories. Sir Francis showed me the telegrams he had received about the German check and retreat in France; and Prince Koudacheff, the Russian Minister, who joined us for coffee, vied with him by showing me his telegrams about the Russian advance in Eastern Prussia and in Austria.

After luncheon, I had some pow-wows on the subject that had brought me, and went to see various people for whom I had messages. They are a lot more cheerful than the last time I was in Antwerp, and are ready for anything.

From the Foreign Office, I went to the Consulate General, where I found a mountain of letters and telegrams. Got off my cables, and answered as much of the other correspondence as was absolutely necessary—no more.

On my way back to the hotel, I ran into General Jungbluth coming out of the Palace, and was promptly hauled inside for gossip.

The Queen, who has very properly come back from England, walked in on us and stopped to hear the news from Brussels.

I got back to the hotel, and found all the colleagues waiting for me to hear the latest news from Brussels. I played my part, and was nearly torn to pieces in their eagerness for news from the town where there is none. They were all there except the Papal Nuncio, who is most unhappy in the midst of war's alarms and hardly budges from the episcopal palace.

After dinner I was again asked to go to the Grand Hôtel to see the Prime Minister. He had nothing startling to say, but was anxious to know what was going on in Brussels. He showed me his telegrams from France, England and Russia, and his maps with the recent movements worked out with little flags.

Monsieur de Brocqueville told me an interesting incident that had taken place at Ghent. It seems that when the Germans arrived there, they sent in an officer and several soldiers to arrange for requisitions, etc., a promise having been given that they would not be molested. Of course, the whole town was on the *qui vive*, and everybody had been warned to refrain from incurring their displeasure. Just as the German motor passed in front of our Consulate, a Belgian armoured car came charging in from Antwerp, knowing nothing of the presence of the Germans, and upon seeing the enemy uniform, opened fire, wounding the officer and one of the men.

That was enough to start things, and the town would probably be in ruins to-day but for the quick thinking and action of Van Hee, the American Vice-

Consul. He plunged down the staircase, seized the Burgomaster, who happened to be present, pushed him into a motor with the wounded men and went straight to the German headquarters to explain that the attack had been made by two men from Antwerp who knew nothing of the agreement reached between the city and the German forces, and to plead that no reprisals should be made upon the city. The general said that he was prepared to accept the statement of the Vice-Consul on this matter, and that he would not therefore visit retribution on the town if the requisitions which he had demanded were promptly furnished. The requisitions were heavy, and he was apparently afraid that they might not be sent. He said that he would send in troops to occupy the town until the supplies requisitioned were actually in his possession, but finally agreed to refrain from doing so on condition that the Vice-Consul should give his word of honour that the supplies should be forthcoming.

Van Hee took this responsibility, and the General agreed to keep his troops outside the town. When they got back to Ghent, the Military Governor disavowed the arrangement on the ground that the Burgomaster had no right to enter into an agreement with the Germans and that he, as Military Governor, was the only one with any authority to deal with them. He therefore declared that no supplies should be sent. The Burgomaster telegraphed the Prime Minister in Antwerp, and placed the entire situation before him, and Monsieur de Brocqueville promptly telegraphed back that since the American Vice-Consul had given his word of honour to the German General it was impossible to disavow the agreement, and that the supplies should be sent out immediately. This was a pretty high stand for the Belgians to take, but they feel that Van Hee saved Ghent from destruction, and are correspondingly grateful to him.

Getting around Antwerp in the evening is quite an undertaking at this time; no street lamps are lighted, all the window shades lined with black, and heavy black shades are placed over the small electric lights in the courtyards of hotels, etc.—all of this to keep from giving any indication to the Zeppelins as to where to drop their visiting cards. A heavy detachment of soldiers guards the approach to the Saint Antoine, and there are patrols in all the streets. The few motors allowed on the street have no lights, and are stopped by all the patrols, who do not call out but rise up silently in front of you and demand the password. It is a ticklish business finding one's way. The big searchlights on the forts sweep the skies from nightfall until dawn, making a wonderful sort of fireworks.

When I got back to the hotel I found Prince Caraman Chimay waiting for me with a message from the Queen. Also poor Prince Ernest de Ligne, whose son, Badouin, was killed in one of the armoured motors several days ago.

Young de Ligne, who was a volunteer, was in one of three armoured cars that went out on a reconnaissance toward the German lines. Just before entering a sunken road between two fields they stopped a Flemish peasant and asked him whether there were any Germans anywhere about. The peasant told them that three Uhlans had been seen a short time before but they had gone away. The three motors, de Ligne in the first, started down and were attacked by about forty Germans under command of a major. De Ligne was shot in the head and died shortly afterwards. The man who took his place at the wheel was killed, and several others of the party were also badly wounded and have since died. The third motor came up from some little distance behind and opened on the Germans, killing or wounding nearly all of them, including the officer, who was killed.

A young chap named Strauss, whose mother was an American, had the mitrailleuse in his car, and stood upright, firing upon the Germans without being touched by the heavy rifle fire that they directed against him. When the Germans had been put to flight he and the other survivors got the three cars into running order, and brought them all back to Antwerp, where de Ligne and two of the others died.

Prince Ernest had a hard time getting through from Brussels, and was fired on several times by the German troops, who were even more nervous than in the morning. when I came through. One of his nephews has also been killed, and another nephew, Prince Henri de Ligne, is in the aviation corps, and has been in the thick of it ever since the beginning of the war. He and his wife are also staying at the Saint Antoine.

On Thursday morning I got caught in another avalanche of telegrams and had to spend a couple of hours at the Consulate-General polishing off and finishing business. Stopped in at the palace on the way back and saw General Jungbluth, who showed me the latest telegrams. I gathered up what newspapers I could beg or buy and stuffed them into a military pouch to take back. Had an early lunch, gathered up M. de Woeste and Faura, whom I was to bring back, and started about one. We got through Malines, across the only one of the three bridges which is left, and started down the bank of the canal toward Hofstade, where Herwarth was to meet us at two o'clock. There was heavy firing by small guns ahead and a certain amount of protective firing from the forts behind us, with the shells singing high above our heads, but we thought that it was probably aimed further to the south and that we could get through.

Just at the edge of Malines we were startled by a tremendous report near-by, and on getting out to reconnoitre I discovered a Belgian battery, which had been established near the Convent of the Dames de Coloma. The commanding officer of the battery, Major Nyssens, whom I had known in Brussels, advised us to wait a little to see if there was a lull in the fighting, so that we would get through. We went into the convent to wait and were warmly received by a little Irish nun, who showed us the park and pictures by way of entertainment, although we felt a much greater interest in the banging of the battery. After a bit Major Nyssens sent out a messenger to the farthest battery to see whether they were prepared to stop firing for a little while to let us scuttle through to Hofstade. Presently an answer came back that at 2:10 the firing would be stopped for twelve minutes to let us through. We were in the motor ready to start when another messenger came from the outer battery saying that the Germans were prepared to move up their battery from the bridge at Hofstade—the very spot we were making for, if there were any lull in the firing and that the Belgian battery could not stop without endangering its position.

We then decided to go back to Malines and to try a direct road by way of Sempst and Villevorde. On parting I gave Nyssens all my cigars, knowing I should find plenty when I got back to Brussels, and he, in a burst of gratitude, gave me a tiny revolver taken off a dead German officer a few hours before. Immediately after getting the revolver Nyssens' orderly had handled it rather carelessly, and shot himself in the stomach. To make sure of doing nothing equally foolish, I took out the remaining cartridges and chucked them in the canal as we rode back to Malines.

About a kilometer out of Malines we ran into a considerable detachment of Belgian infantry and lancers and a large armoured motor with two mitrailleuses. We were told that the Belgians had taken and retaken Sempst three times during the day, and while neither side occupied the town at that precise moment they were both advancing on it, and that it might be rather warm for ordinary motors. They finally agreed to let us talk to the commanding officer, who turned out to be none other than Colonel Cumont, the owner of the building occupied by the Legation.

He was up on a railway embankment, lying on his stomach between the rails, watching some German patrols through a pair of big field glasses, and when we hailed him, rolled gracefully over the side, and came down to talk to us. He had been out on the track most of the time for three days and was a rather disreputable-looking person, but apparently glad of a chance to talk with someone from the outside world.

He said he thought we would have time to get through before the row began, and in any event he would warn his men so that if we came scuttling back we would be given the right of way to safety.

We passed several Belgian patrols along the way and finally got into the town, which showed clear evidences of fighting; some of the houses were burned to the ground, and all that were standing had their doors and windows smashed, furniture broken, and strewn about the floors with broken bottles and dishes, mattresses and goodness knows what else; and above all arose that terrible smell of burnt flesh.

We were nearly through the town when we were hailed by a detachment of about twenty Belgians, who had got through and occupied the grounds of a villa on the edge of the village. We stopped the car, and I got out and went ahead, they remaining with leveled rifles, in their usual hospitable manner. When I got to within twenty feet of them we heard the whirr of a machine gun—which the Belgian soldiers call a *cinema* —and a German armoured car poked its nose around the corner for a look-see. It was firing high to draw a return fire and locate any Belgians there might be in the town, but they all scurried behind cover, closely followed by me. They were taking no chances, however, and called me to stay in the middle of the road. Without wasting any time in formality I made clear my identity, and, on being shown through a breach in the wall a disagreeable-looking body of German infantry and lancers about a half a mile away approaching through a field, I decided that we were on the wrong road and made back for the motor.

I told my passengers what was up, and that we had to go back to Malines. M. de Woeste, however, was all for going through on the valid plea that he had no clean linen and did not want to spend another night out of Brussels. Nevertheless we turned around and started back, only to rush into the big Belgian armoured car which Colonel Cumont, hearing firing, had sent down to rescue us and cover our retreat. This car stayed in the village for a few minutes to meet the German car, fired a few shots at it, and then came back to the outposts.

We then tried getting out toward the west from Malines, but soon came to a point where the road was inundated, and had to turn back for the third time. It was then getting pretty late in the afternoon, and even M. de Woeste had to admit that we had best come back to Antwerp rather than try to make a roundabout journey to Brussels after dark.

All the way back into Antwerp we met Belgian forces advancing to the attack. They are getting to know the flag better every day and we were greeted with waving hands and cheers everywhere we went. When nearly in town, a young chap ran out of the ranks to where we were waiting for them to get by, grabbed me by the shoulder, and said:

"I am born an American."

"Where were you born?"

"Aurora, Illinois. My father worked in ———'s glycerine works."

"Who do you know in Aurora?"

"I know Mr. Evans and Mr.—— and Mr. ——
and Mr. *Beaupré.*" *

"What's your name?"!

Just then a non-commissioned officer came along and
ordered him back into the ranks; the motor started
ahead, and I lost track of the boy in a cloud of dust.

At the edge of town we caught up with a British
Legation motor, which was stopped at a railroad barricade. Its occupants roared with laughter when they
saw us, and Colonel Fairholme gloated particularly,
as he had prophesied that we would not get through.
When we got back to the hotel we were met with more
laughter. It was the great joke of the week to see the
only people who had previously been successful in running the lines, caught like the rest of them. I was not
at all down in the mouth, as Antwerp was most interesting, and I had left only because I had felt it my
duty to get back to work and to keep the Minister
from worrying. When I saw that there was no way
of getting through I gladly accepted the decree of fate.

When we got back to Antwerp I soon learned that
it would be out of the question to get back to Brussels
the next day, or perhaps even the day after that. The
Belgians were advancing to an enveloping movement
and all the surrounding country was to be covered with
Belgian troops in an endeavour to deal a smashing blow
to the Germans and compel them to bring back more
troops from the front in France. Colonel Fairholme
asked me to accompany him to the front next morning, and I accepted with an alacrity which startled him.

---

* Former American Minister at The Hague.

After dinner I made another excursion into the darkness and told Monsieur de Woeste that there was no prospect of getting back to Brussels the next day. His colleagues, who were there also, impressed upon him the futility of going, and he finally resigned himself to staying, although he kept insisting that he infinitely preferred danger to boredom, which was his lot so long, as he had nothing to do but sit around the hotel.

Friday morning while I was waiting for the Colonel to get ready and was doing my little errands down town, there came a great roaring of a crowd, and the chauffeur, knowing my curiosity, put on steam and spurted down to the boulevards just in time to run into a batch of three hundred German prisoners being brought in. They were a dejected-looking crowd, most of them Landsturm, haggard and sullen. The crowd, mindful of the things the Germans have been doing to this little country, were in no friendly mood, but did nothing violent. There was only a small guard of Belgian Garde Civique to escort the prisoners, but there were no brickbats or vegetables. The people limited themselves to hoots and catcalls and hisses— which were pretty thick. And even this was frowned upon by the authorities. Within a couple of hours the Military Governor had posted a proclamation begging the people of Antwerp to maintain a more dignified attitude and to refrain from any hostile demonstration against other prisoners. This batch was surrounded, and caught at Aerschot, where the Germans are said to have committed all sorts of atrocities for the past three weeks. Among the prisoners was the commanding officer, who was accused of being responsible for a lot of the outrages. He was examined by the military court, which sits for the purpose, and admitted having done most of the things of which he was accused, pleading in his own defence that he had done them only in obedience to superior orders, to which he had protested. The soldiers who made the capture disclaimed a large part of the credit for it on the ground that most of the Germans were drunk and that they were too dazed to get to their arms. Stories of this sort keep piling in from every side.

We got away at eleven to Lierre, where the King has established his headquarters for his movement. The road lay to the southeast and was through country I had not traversed before. The aspect was the same, however—long stretches of destroyed houses and felled trees, barbed-wire entanglements and inundated fields. It is a mournful sight.

Little Lierre was unharmed, and I hope it may remain so. The Grande Place was filled with staff motors, and there was a constant coming and going of motors and motorcycles bearing messengers to and from the field of operations. Headquarters was established in the Hôtel de Ville, which bears on its tower the date 1369—a fine old building, not large, but beautiful.

In the morning a message had come ordering Colonel DuCane back to England. He was out in the field, and we had to wait until he came in to deliver it to him. The King was also away, but we put in our time talking with the officers on duty as to the movement and its progress, and then went out for a stroll around the town. We looked into the old church, and I stopped and bought an officer's forage cap as a souvenir of the place. By the time we had poked around the neighbourhood and inspected the other *Sehenswürdigkeiten* of the town it was lunch time and we joined an officers' mess in the back room of a little café on the square, and then, to kill time, sat in front of another café and had coffee and a cigar.

We could not get started until Colonel DuCane had returned and received his message, so we sat in front of our little café and growled. It was maddening to waste our time there while the guns were thundering all around us and we knew from the signs of activity at headquarters that big things were toward. After a time a little man, the Senator for the district, came out and asked us into his house, directly across the street from the Hôtel de Ville. It was raining hard and we were ready for a change, so we accepted gladly and were entertained with champagne and cigars to the music of falling rain and booming cannon.

Our Senator was very much down in the mouth about the situation in general and wanted to talk about it. The Colonel told him of the bulletins that had been published in. Antwerp as to the progress of the campaign, and as this went on he cheered up visibly minute by minute—whether as a result of the good news or the champagne, I don't know.

The Colonel was called away after a time to talk to Lord Kitchener over the telephone. Kitchener keeps himself informed directly as to the progress of operations and the knowledge that he may drop in over the telephone at any minute gives his officers a very comforting feeling that they are not forgotten

Finally, after dark, Colonel DuCane and Captain Ferguson came in, and we got under way. It was too late to go forward with hopes of seeing anything, but it was evident that things would be as hot as ever the next day and that I could not hope to get my charges back to Brussels. Accordingly the Colonel's invitation was extended and accepted, and we turned back toward Antwerp considerably disappointed.

While we were waiting around trying to make up our minds—if any—I ran into young Strauss, the half-American, who was in the armoured car behind young de Ligne. He was really the principal hero of the occasion, having stood bolt upright in his car and riddled the German forces with his mitrailleuse until the few survivors turned and fled. He had with him two of the other survivors of his party. All of them had been decorated with the Order of Leopold for their behaviour. An order like that looks pretty well on a private's uniform, particularly when given with such good reason.

We had retreated inside the Hôtel de Ville during a particularly heavy downpour of rain, when in came the King, who had spent the whole day in the field with the troops. He was drenched to the skin, but came briskly up the steps, talking seriously with his aide-de-camp. He stopped and spoke with us all and took Colonel DuCane into his study and had a few minutes talk with him by way of farewell. The King shows up finely in the present situation and all the foreign military attachés are enthusiastic about his ability. He is in supreme command of the army and no detail is too insignificant for his attention.

At Malines—a good background for a photograph to send home to Germany

His Eminence, Cardinal Mercier, Archbishop of Malines

We got the password and made back for Antwerp in the dark, leaving Colonel DuCane and Captain Ferguson to spend the night at Lierre. We were in bad luck and got stopped at every railroad crossing along the way. Troop and supply trains were pouring down toward the front and Red Cross trains were bringing back the wounded in large numbers. Both sides must have suffered heavily during the day, and there may be several days more of this sort of fighting before there is a lull.

When we got back to the hotel we found Sir Francis waiting for us with a glowing telegram and an equally glowing face. It was the most enthusiastic message yet received from the British War Office, which has been very restrained in its daily bulletins. For the first time that day it spoke with a little punch, speaking of the "routed enemy" and their being "vigorously pressed." We tumbled through a hasty bath and got down to dinner in short order.

After dinner it was the same old performance of going over to the Grand Hôtel and labouring with Monsieur de Woeste, who was still bent on getting home to his clean linen without further delay. It took the united arguments of the Cabinet, which was in session, to convince him that it would be useless and foolish to try to get away. Finally he yielded, with a worse grace than on the previous evening. I had a comfortable visit with several of the Ministers, who were glad to hear news of their families in Brussels, and asked me to remember all sorts of messages to be given on my return. I only hope that I shall not get the messages mixed and get too affectionate with the wrong people.

The Cabinet was going through the latest telegrams from the various fields of action. They even had some from Servia and were decidedly cheered up, a big change from the dogged determination with which they were facing bad news the last time I was in Antwerp.

Saturday morning the Colonel and I were called at six, and at seven we got away in a pouring rain over the same road to Lierre that we had travelled the day before. There was a big force of workmen hard at it in the vicinity of the outer forts, burning houses and chopping down trees and building barbed-wire entanglements. It is a scene of desolation, but it is necessary in a fight like this.

We found things moving rapidly at headquarters in Lierre. Messengers were pouring in and orders going out with twice the activity of the day before. The movement had been under way for two hours when we got there and the guns were booming all around. After learning as much as we could of the disposition of the troops we went out and stocked up with bread, cheese, and mineral water, and started forth to see what we could of the operations. We took along a young officer from headquarters to show us the road. We soon saw that he did not know the roads and could not even read a map, and had to take over that work ourselves. Colonel Fairholme and I went in my motor with the headquarters passenger and Colonel DuCane and Ferguson followed in their own car with an orderly. We got to Malines without difficulty and got out for a look at the Cathedral. It is a dreadful sight, all the wonderful old fifteenth century glass in powder on the floor. Part of the roof is caved in and there are great gaping holes in the lawn, showing where the shells struck that fell short of their mark. A few of the surrounding houses, belonging to entirely peaceful citizens, were completely wiped out while they were getting the range. It is hard to see what useful military purpose is served by smashing churches and peaceful habitations, when there are no troops about the place. Malines was bombarded when the troops had withdrawn. It is hard to reconcile with *Gott mit uns*.

Before we left Lierre, nine troopers of the Landsturm were marched into the hallway of the Hôtel de Ville, to be examined by the officer who is there for that purpose. They were a depressed lot who had run away and given themselves up, so as to be spared the hardships and dangers of the rest of the war. They answered questions freely, telling all they knew as to the disposition of troops and making their getaway toward the local lockup with great alacrity as soon as the word was given to move. Most of them were Bavarians. Colonel Fairholme speaks German like a native. He talked with these chaps, and there was some interesting conversation. They were all without enthusiasm for the war, and all expressed indignation at having been brought out of the country, maintaining that the Landsturm cannot be used for anything except the maintenance of order in the Empire. I think they are wrong about that, but this was no joint debate on German law, and no attempt was made to sooth their injured feelings. A lot of men were brought in while we were there, some of them prisoners taken during the fighting, but a great many of them fugitives who were sick of the war, and only asked to get off with a whole skin.

As they marched out of the hall, the King came in from the field for a look at the morning's telegrams. He had been out since long before daybreak, and was covered with rain and mud. He shook himself vigorously, spraying everybody with raindrops, and then stopped to speak to us before going in for a cup of coffee and a look at the news.

From Malines we made back along the northern side of the canal, in an endeavour to find the headquarters of the —th Division. We went through a little village where all the inhabitants were standing in the road, listening to the cannonading, and spun out upon an empty and suspiciously silent country road. A little way out we found a couple of dead horses which the thrifty peasants had already got out and skinned. I didn't like the looks of it, and in a minute the Colonel agreed that he thought it did not look like a road behind the lines, but our little staff officer was cocksure that he knew just what he was talking about, and ordered the chauffeur to go ahead. Then we heard three sharp toots on the horn of the car behind—the signal to stop and wait. And it came pulling up alongside with an inquiry as to what we meant by "barging" along this sort of a road which likely as not would land us straight inside the enemy's lines. There was a spirited discussion as to whether we should go ahead or go back and strike over through Rymenam, when we heard a shell burst over the road about half a mile ahead, and then saw a motor filled with Belgian soldiers coming back toward us full tilt. The Colonel stopped them and learned that they had been out on a reconnaissance with a motor-cyclist to locate the

The children of Antwerp played at soldiering through the siege

The nuns, scornful of danger, stayed where they could render the
greatest service

German troops in front of Hôtel de Ville, Brussels

Types of von Arnim's troops

German lines, which were found to be just beyond where the shell had burst, killing the motor-cyclist. It would have been a little too ignominious for us to have gone bowling straight into the lines and get taken prisoners. We turned around and left that road to return no more that way. We got about half-way up to Rymenam when we met some Belgian officers in a motor, who told us that a battery of the big French howitzers, which had just gone into action for the first time, were in a wood near H——. We turned around once more, and made for H—— by way of Malines. We found the headquarters of the —th Division, and went in and watched the news come in over the field telephone and telegraph, and by messengers on motor-cycles, bicycles and horses straight from the field. The headquarters was established in a little roadside inn about half a mile outside the town, and was as orderly as a bank. Officers sat at the various instruments and took notes of the different reports as they came in. Reports were discussed quickly but quietly, and orders sent out promptly but without confusion. The maps were kept up to the minute by changing the little flags to show the positions of the different troops right at the minute. There was telephone communication with the forts, and several times they were ordered to pour fire into a certain spot to cover an advance or a retreat of parts of the Belgian forces, and, at other times, to cease firing, so as to let Belgian troops cross or occupy the exact spot they had been bombarding. It was a wonderful sight to watch, and it was hard to realise that this was merely a highly scientific business of killing human beings on a large scale. It was so business-like and without animus, that to anyone not knowing the language or conditions, it might have passed as a busy day in a war office commissary when ordering supplies and giving orders for shipment.

Just outside the headquarters was one of the fine German kitchen wagons with two fine Norman horses which had pulled it all the way from Germany. It had been stationed in the grounds of a château not far away, and three men of its crew were hard at work getting a meal when a little Belgian soldier with two weeks' growth of beard waltzed into the garden, shot one of the men dead and captured the other two. He disarmed them, put ropes around their necks and drove the kitchen to headquarters in triumph. He was proud as punch of his exploit, and, for that matter, so was everybody else around the place.

In a field of turnips a couple of hundred yards away from the headquarters were the howitzers. There were three of them in a row with three ammunition wagons. They had been sent here only a few days ago, and they were promptly put into action. They were planted here, slightly inside the range of the guns from the outer forts, and were able to drop shells six miles from where we stood, or about five miles outside the range of the fort guns. They toss a shell about two feet long, filled with deadly white powder, six miles in ten seconds, and when the shell strikes anything, "it thoes rocks at yeh!" as the darkey said about our navy guns. The battery was planted down behind a little clump of pines, and was dropping shells into a little village where there was a considerable force of Germans about to be attacked. The Germans must have been puzzled by this development, for they had counted on being able to advance safely up to the range of the forts, feeling sure that the Belgians had no powerful field guns of this sort.

We were introduced to the officers commanding the battery, and watched their work for nearly two hours. One of the officers was Count Guy d'Oultremont, adjutant of the Court, whom I had known in Brussels. He was brown as a berry, had lost a lot of superfluous flesh, and was really a fine-looking man. He had been in Namur, and had got away with the Belgian troops who went out the back door into France and came home by ship.

After we had been watching a little while, an aeroplane came circling around, evidently to spot the place where these deadly cannon were. It cruised around for some time in vain, but finally crossed straight overhead. As soon as we were located, the machine darted away to spread the news, so that the big German guns could be trained on us and silence the battery; but the Belgians were Johnny-at-the-rathole again, and he was winged by rifle fire from a crowd of soldiers who were resting near the headquarters. They killed the observer and wounded the pilot himself, to say nothing of poking a hole in the oil tank. The machine volplaned to earth a few hundred yards from where we were, and the pilot was made prisoner. The machine was hauled back to the village and shipped on the first outgoing train to Antwerp as a trophy.

We were leaving the battery and were slipping and

sliding through the cabbages on our way back to the road, when we met the King on foot, accompanied only by an aide-de-camp, coming in for a look at the big guns. He stopped and spoke to us and finally settled down for a real talk, evidently thinking that this was as good a time as any other he was likely to find in the immediate future.

After talking shop with the two colonels, he turned to me for the latest gossip. He asked me about the story that the German officers had drunk his wine at the Palace in Laeken. I told him that it was generally accepted in Brussels, and gave him my authority for the yarn. He chuckled a little and then said, in his quiet way, with a merry twinkle: "You know I never drink anything but water." He cogitated a minute and then, with an increased twinkle, he added: "And it was not very good wine!" He seemed to think that he had quite a joke on the Germans.

As we talked, the sound of firing came from the German lines not far away, and shrapnel began falling in a field on the other side of the road. The Germans were evidently trying to locate the battery in that way. Most of the shrapnel burst in the air and did no damage, but some of it fell to the ground before bursting and sent up great fountains of the soft black earth with a cloud of gray smoke with murky yellow splotches in it. It was not a reassuring sight, and I was perfectly willing to go away from there, but being a true diplomat, I remembered that the King ranked me by several degrees in the hierarchy, and that he must give the sign of departure. Kings seem powerless to move at such times, however, so we stayed and talked while the nasty things popped. His Majesty and I climbed to a dignified position on a pile of rubbish, whence we could get a good view up and down the road, and see the French guns which were in action again.

A little later Ferguson, who was standing not far away, got hit with a little sliver and had a hole punched in the shoulder of his overcoat. It stopped there, however, and did not hurt him in the least. He looked rather astonished, pulled the little stranger from the hole it had made, looked at it quizzically, and then put it in his pocket and went on watching the French guns. I think he would have been quite justified in stopping the battle and showing his trophy to everybody on both sides.

The King was much interested in all the news from Brussels, how the people were behaving, what the Germans were doing, whether there were crowds on the streets, and how the town felt about the performances of the army.

He realised what has happened to his little country, and made me realise it for the first time. He said that France was having a hard time, but added that perhaps a sixth of her territory was invaded and occupied, but that every bit of his country had been ravaged and devastated with the exception of the little bit by the sea coast and Antwerp itself, which was getting pretty rough treatment, in order to put it in shape to defend itself. He spoke with a great deal of feeling. And no wonder!

Then to change the tone of the conversation, he

looked down at my pretty patent leather shoes, and asked in a bantering way whether those were a part of my fighting kit, and where I had got them. I answered: "I got them several months ago to make my first bow to Your Majesty, at Laeken!" He looked around for a bit at the soggy fields, the marching troops, and then down at the steaming manure heap, and remarked with a little quirk to his lips: "We did not think then that we should hold our first good conversation in a place like this, did we?" He smiled in a sad way, but there was a lot more sadness than mirth in what he said.

Guy d'Oultremont came up and said something that I did not understand, and we started back toward the headquarters. We stopped opposite the inn, and the two colonels were called up for a little more talk.

Just then a crowd of priests, with Red Cross brassards on their arms, came down the road on their way to the battlefield to gather up the wounded. With his usual shyness the King withdrew a few steps to seek shelter behind a motor that was standing near by. As we talked, we edged back a little, forcing him to come forward, so that he was in plain sight of the priests, who promptly broke out in a hearty *"Vive le roi!"* He blushed and waved his hand at them, and, after they had passed by, shook hands with us and followed them on foot out onto the field. In modern warfare a King's place is supposed to be in a perfectly safe spot, well back of the firing line, but he does not play the game that way. Every day since the war began, he has gone straight out into the thick of it, with the shells bursting all around and even within range of hostile rifle fire. It is a dangerous thing for him to do, but it does the troops good, and puts heart into them for the desperate fighting they are called upon to do. They are all splendidly devoted to him.

The rain stopped as we got into the motors and started back toward Malines, with the idea of locating the other battery of *obusiers*. There was a sharp volley of three toots on Colonel DuCane's horn, and we came to a sudden stop, with the emergency brakes on, to receive the information that it was two o'clock and time for lunch. None of us had kept any track of time, and all were ready to go sailing along indefinitely without food. As soon as we had noticed the time, however, we all became instantly hungry, and moved along, looking for a good place for lunch. I had the happy idea of suggesting the convent where we had taken refuge on Thursday, and thither we repaired to be most warmly greeted by all the nuns, and most particularly by the little Irish sister who was overjoyed to see British uniforms and hear some war news that she could believe. She hailed me with, "Oh! and it's the riprisintitive of the Prisidint!" The nuns gave us a table in the park and two big benches, and we got out our bread and cheese and chocolate and a few other things that Colonel DuCane had found somewhere, and had a most comfortable meal with a towering pitcher of beer brought out from the convent, to give us valour for the afternoon's work.

After lunch we went back through Malines again, through the railroad yards, bumping over the tracks, and away toward Muysen and Rymenam to see the other batteries. I was struck in going through the railway yards, which I had always seen teeming with activity and movement, to see that all the rails are covered deep with rust—probably for the first time. Think of it!

After leaving Muysen, our road lay for a mile or so along a canal with open fields on either side. Uhlan patrols had been reported in this part of the country, which was in a weak spot in the Belgian lines, and the Colonel told the staff officer to keep a sharp lookout and be ready with his revolver and prepared for a burst of speed. That military genius replied with an air of assurance: "Oh, that's all right. They cannot cross the canal." The Colonel confined himself to saying mildly: "No, but bullets can!" Little Napoleon said nothing more, but I noticed that he unstrapped his revolver without loss of time.

We were bowling along the road, looking for the battery, when there was the most enormous noise which tore the earth asunder and the universe trembled. I looked around to the left, and there not more than a hundred feet away were those three husky French guns which had just gone off right over our heads! We had found them all right, but I should prefer to find them in some other way next time.

We spent a little time looking at them, and Ferguson had them get out some of the explosive and show it to me. It comes in long strips that look for all the world like chewing gum—the strips about the same proportions, only longer. I fail to see, however, how they can be made to blow up.

After a bit we got back into the cars, and started

The Hôtel de Ville, Louvain

Belgian War Medals

out to cruise around to the Belgian left wing and watch a little of the infantry fighting at close quarters. We very soon began running into stragglers who informed us that the —th Division was being driven back, and that a retreat was in progress. Soon we came upon supply trains and ammunition wagons making for the rear, to be out of the way of the troops when they began to move. We were not anxious to be tangled up in the midst of a retreat, and obliged to spend the night trying to work our way out of it, so we forged ahead and got back to Lierre as fast as we could. It was raining hard as we came in, and we took refuge in the Hôtel de Ville, where the colonels read their telegrams and got off a report to London. One of their telegrams brought the unwelcome news that Ferguson was also recalled to England. They are evidently hard put to it to find enough officers to handle the volunteer forces. He will have to stay on for a few days, but Colonel DuCane came back with us and left the next morning for England by way of Ostend.

When we got back to the hotel after a fast run, I found that Inglebleek, the King's Secretary, had been around twice for me, and wanted me to go at once to the Palace. I jumped into the car and ran over there, to learn that the Queen wanted to see me. She was then at dinner, and he thought it would do the next time I came up—she seems to have wanted more news of Brussels—nothing pressing. She had told Inglebleek to give me a set of the pictures she had had taken of the damage done to the Cathedral at Malines. They are interesting as a matter of record.

Sir Francis had another good bulletin from the War Office, and was beaming. The colleagues came and gathered round the table, and chortled with satisfaction.

Heavy cannonading continued well into the night, to cover the advance of the —th Division, which had been reinforced and was moving back into the dark and rain to take up its old position and be ready for the Germans in the morning.

I was up and about early on Sunday morning. Had breakfast with Count Goblet d'Alviella, one of the Ministers of State. Gathered up Monsieur de Woeste and Faura, and made for the Scheldt and Brussels. Instead of going across on the boat as we had to do the last time, we found a broad and comfortable pontoon bridge placed on canal boats and schooners lashed together and moored from one side of the river to the other. Any time they like, the Belgians can cut the string, and there is no way of getting into the city from that side. There was a tremendous wind blowing and the rain fell in torrents—short showers—from the time we left Antwerp until we came sailing into town here.

The bridge at Termonde had been blown up by the Germans on evacuating the place after having destroyed the entire town, so there was no thought of returning that way. I knew there could be nothing doing the direct way through Malines, so decided on a long swing around the circle by way of Ghent as the only practicable way. We found Belgian troops all the way to Ghent, and had no trouble beyond giving the password which I had. We drew up at a restaurant in a downpour and had a hasty lunch, getting under way again immediately afterward.

About ten kilometers this side of Ghent we came to Melle, a village which had been destroyed, and another where a number of houses had been burned. A nice-looking young chap told us that there had been a fight there the day before and that the Germans had set fire to the place as they retreated—just from cussedness, so far as he could see. There, and at another place along the road, peasants told us that they had been made to march in front of the German troops when they marched against the Belgians. I don't like to believe that there is any truth in that story but it comes from every direction and the people tell it in a most convincing way.

We found no Germans until we were this side of Assche and then our adventures were evidently at an end. As we came in we could hear heavy cannonading from the direction of Vilvorde and Hofstade and knew that the fight was still going on. They had been hearing it in town for a couple of days.

The family at the Legation had been somewhat anxious, but had learned through the Germans that we were all right—evidently from somebody who got through the lines. I had to sit right down and tell the story of my life from one end to the other.

I never got over the idea in Antwerp of the incongruity of going out onto the field all day and fighting a big battle, or rather, watching it fought, and then sailing comfortably home to a big modern hotel in a motor and dressing for dinner. I don't think there has ever been a war quite like this before.

Herwarth has gone to the front for some active service. I am sorry to miss him. He went up to Hofstade the day I was to have returned, and waited for me about an hour, but the fire got too thick for him and he came back and reported that I would not be able to get through.

Monsieur de Woeste called this afternoon and paid his respects. He gave the Minister an account of the attempts we made to get through that made his hair stand on end for an hour afterward.

Brussels, September 16, 1914.—To-day has brought a long string of callers, and between times we took satisfying looks at the passing troops, which have been pouring into town steadily yesterday and to-day. Nobody has established to my satisfaction whence they come or whither they are going. There are all sorts of explanations offered, each explanation being quite convincing to the one who offers it. Most people say that they are being brought in for the siege of Antwerp, which is about to begin. The siege of Antwerp has begun so often and never materialized that I decline to get excited about it at this stage of the game. Another explanation is that the German retreat in France is so precipitate that some of the troops and supply trains are already pouring through here on their way home. I cannot get up much enthusiasm for that either. Some imaginative souls maintain that these are forces being brought back to fight against the Russians. None of these stories sound good to me and I have resigned myself to the belief that the only really safe conjecture is that this "is a movement of troops."

This morning Baron von der Lancken came in and asked me to testify as to what we had seen at Louvain. Of course what we saw had no bearing on the original cause of the trouble and there is no reason for me to push my way into the controversy. Besides, I can't do it without orders from Washington.

We are getting quite accustomed to having no communications with the outside world. Railroads, of course, have ceased to work, except for military purposes, and there is no way for the general public to get about. There has been no postal service since the Germans marched in on August 20th, and we don't know when we shall have any. All telephones were cut off within a few hours of the arrival of the German army. There are no newspapers, and all the information we are supposed to have about happenings in the outside world is fed to us in the form of placards on the walls of the city. Nobody takes any great amount of stock in what these placards tell us, although they have sometimes told us the truth, and consequently there is a great demand for the few copies of Dutch and English newspapers that are smuggled across the border and brought to Brussels. The prices vary according to the number of papers to be had, and run from five francs to one hundred francs for a single copy of the *Times*. Those who do not care to spend so much can rent a paper by the hour—and customers are not wanting on this basis. By way of discouraging this traffic it is said that the Germans have shot several men caught smuggling papers. Those caught selling them in Brussels are arrested and given stiff terms of imprisonment. All taxis disappeared many days ago and altogether the normal life of the town has ceased. It will be a rollicking place from now on.

Brussels, September 17, 1914.—This morning I spent digging my way out from under a landslide of detail work which has been piling up on my desk, until I could hardly see over it. I now have it out of the way, and can breathe again freely for the moment.

This afternoon Baron de Menten de Horne, a Lieutenant in the Second Regiment of Lancers, was brought in to the Legation, a prisoner, still wearing his Belgian uniform. He was captured last Friday near H—— while I was there. Nyssens, the Major who was in the convent with us, told me that one of his officers had gone off on a reconnaissance and had not reappeared; he was greatly worried about him, but could not send any one out to look for him. This was the man. He was surrounded, in company with several of his men, and took to cover in a field of beets. Night was coming on, and they thought that when the fight was over and the German troops who were all about them had retired, they would be able to work their way out and rejoin their own forces, but twenty-five Germans surrounded them, and after killing all the others, took this man prisoner.

His only idea is to be exchanged and rejoin his regiment; and, as is the case with pretty much everybody else nowadays, he turned to the American Legation. He made such a good plea that the German authorities brought him here yesterday, and left him an hour, on his giving his word of honour not to divulge anything as to the military movements he had seen while a prisoner.

Of course, we could not arrange to make the exchange, but he stayed on for an hour and told us of his adventures. He was a pathetic figure in his dirty uniform, sitting on a little chair in my office and telling in a simple way of all he had been through—laying more stress on the sufferings and death of his soldiers than on anything that had happened to him. His own brother had been killed in the fighting around Liège, and he had heard that his brother-in-law, of whom he was very fond, had also been mortally wounded. While at Louvain, he had visited the military hospitals, and had a list of Belgian officers who were there. I took a list of them, by permission of the German officer who came after the prisoner, and shall send word to their families.

I went around to see the young man's sister, and sent her off to have a look at him at headquarters, where he is being well treated. It is a joy to be able to do some of these little errands. Nobody can realize the amount of bitter sorrow there is in this country— we cannot realize it ourselves, but now and then a wave of it rises up to confront and overwhelm us.

Miss T——, an American owning a school here, was in late this afternoon to complain of the behaviour of a couple of officers and gentlemen who did her the honour of calling upon her. They came swaggering in, asked whether a certain German girl had attended the school and demanded her portrait. On being refused, they became nasty and finally so overawed the two women who were there alone that they found some snap shots and handed over a couple of them. Then they demanded a post card with a picture of the school, wrote a message to the girl, and tried to compel the two women to sign it. They flatly refused, and, in a rage, the elder German tore up the card, threw it at Miss T——, flung down the photographs and stamped out of the house, slamming the doors.

The Minister is going over to see the military authorities in the morning and make some remarks that they will not forget in a hurry. The puppies ought to be horsewhipped.

September 18th.—Repressive measures are getting stronger and more severe. The Germans have now ordered the Belgians to take down their flags. Lüttwitz, the Military Governor, has posted an *Avis* on the subject which is worth reproducing in full.

The population of Brussels, understanding well its own interests, has generally, since the arrival of the German troops, maintained order and quiet. For this reason, I have not yet forbidden the display of Belgian flags, which is regarded as a provocation by the German troops living in or passing through Brussels. Purely in order to avoid having our troops led to acting on their own initiative, I now call upon houseowners to take down their Belgian flags.

The Military Government, in putting this measure into effect, has not the slightest intention of wounding the susceptibilities and dignity of the citizens. It is intended solely to protect the citizens against harm.

Brussels, September 16, 1914.　　BARON VON LUTTWITZ.

General and Governor.

Dined at the Palace in a din of German officers. Bulle, Pousette and Riseis kept me in countenance. There were also some twenty or thirty Austrian officers—the first we have seen. They were quiet and well behaved, and contrasted sharply with their allies.

Brussels, September 19, 1914.—This morning our Vice-Consul came in from Ghent bringing with him a pouch and a huge bag of letters and telegrams. These had been got through to him from Antwerp yesterday, and he made a run through the lines early this morning, having been turned back several times on account of small engagements between Belgian and German outposts.

This morning a Dutchman came in to see me, and after showing me a lot of papers, to establish that he was somebody entirely different, told me that he was a British spy. He then launched into a long yarn about his travels through the country and the things he had seen, unloading on me a lot of military information or misinformation that he seemed anxious to have me understand. After he had run down I asked why he had honoured me with his confidence, and was somewhat startled to have him answer that he had no way of getting it out and thought that inasmuch as we were charged with the protection of British interests I might have an opportunity to pass it on where it would do the most good. He seemed rather pained at my remarks, and was most reproachful when I threw him out on his head. Yes, my shrewd friend, it has also occurred to me that he may have been a German spy just trying to find out whether we were indulging in dirty work. It would not be the first time that that sort of thing was tried on us.

Monseigneur N—— came around this afternoon and asked me to take him to Antwerp on my next trip. I told him that I could not, as I had already promised to take some other people, and that my car would be full. He said that he had his own car, and that he would ask me to convoy him; he had heard that I had *"beaucoup de bravourr, tandis que moi je n'ai pas de bravourrr et j'aimarais me mettre sous votre protection."* I sent him to see von der Lancken, and he came back in a little while to say that he was told that the only safe way was to go by Namur, Liège and Holland, entering Antwerp from the north. He evidently insisted on a perfectly safe route, that could be guaranteed, and they told him a story that they thought would dissuade him from making the trip. They do not like to have a lot of people coming and going.

We have no more news from the outside world; the battle still rages all along the line in France (according to what we hear), but we have no inkling as to whether the German retreat still continues. The only thing we are told at headquarters is that the outcome is as yet undecided, but that the Germans are in a favourable position, and that they will be victorious in a few days. I would give a good deal for a little real news as to how things are going.

This morning Major Langhorne, our Military Attaché from Berlin, breezed in upon us. He is travelling around with six other Military Attachés, seeing as much of the field of operations as the German officer who personally conducts them will permit. They got in this morning, and left about one, so we had only a few minutes' visit, and he carried off all our good wishes and New York papers.

The German *affiche* of yesterday, ordering the Belgian flags taken down, has made everybody furious, and for a time we thought there might be trouble. If the flags had been ordered down the day the Germans came in there would not have been half as much resentment, but, on the contrary, they began by proclaiming that the patriotic feelings of the people would be scrupulously respected. Max, the Burgomaster, got out a little proclamation of his own which served to soothe the feelings of the people. After expressing some views as to the German order, he says:

I ask the population of the town to give a fresh example of self-restraint and greatness of soul which it has already so often shown during these sad days.

Let us provisionally accept the sacrifice which is imposed upon us; let us take down our flags in order to avoid conflicts, and patiently await the hour of redress.

Soon flags were coming down all over the city, and there was not a murmur. An hour after Max's proclamation was posted, however, German soldiers were running about covering them with sheets of white paper. The Military authorities were furious, because Max had intimated in his poster that the present situation would not endure forever, and that the Belgian flag would fly again over Brussels. In their unimaginative way they sent down a squad of soldiers and arrested him. He was taken to headquarters, and brought before von Lüttwitz, who told him that he was to be taken as a prisoner of war to Berlin. Max replied that he bowed before superior force; that he had done what he knew to be necessary for the preservation of order in his city, and that he was ready to accept the consequences of his act; that at any rate he would have the satisfaction of having maintained order here up to the minute that he was sent to Germany, and that he could not be held responsible for what might happen after his departure. General von Lüttwitz sat up and took notice of the last part of this and rushed off to see von der Goltz. In ten minutes he came back and told Max that he was free and that the Field Marshal desired that he should continue to act as Burgomaster as though nothing had happened. Why don't people have a little imagination!!

The town is still bottled up, and troops are being marched back and forth across it, as, I believe, purely for the purpose of impressing the population with the belief that they are far more numerous than they really are. Late this afternoon I took a drive to the edge of town, and we were stopped half a dozen times and had our papers examined. From all I can gather it would seem that the Germans are entrenching themselves as solidly as they can so as to be ready to resist another sortie without sustaining the terrible losses they suffered last time. They cannot be very happy over the way things have been going in France, although they have this afternoon announced a great victory on their right wing.

One of our friends who has just come back from the coast reports that there were a lot of French troops marching through Belgium on their way from Dunkerque to Lille—evidently an attempt to turn the German right wing. We have heard nothing more about it.

The food supply of the country is being rapidly exhausted and there is urgent need for importations. The public knows little about the situation, but a serious shortage threatens and we must have a considerable stock from abroad. The Brussels committee has raised a goodly sum of money and hopes to get food from Holland and England to meet present needs. Similar committees are being formed in other cities, and they, too, will require food from abroad. The local committee has asked Shaler to go to Holland and from there to England to purchase as much food as possible, make arrangements for sending it across the frontier and investigate the chances of getting future supplies. The German authorities have given assurances that they will not requisition any of the supplies imported for the use of the civil population. They are to issue placards signed by the Military Governor ordering the military authorities to respect our purchases. These placards are to be affixed to the cars and barges bringing in the supplies and we are inclined to believe that they will be effective.

After hurried preparation Shaler got away this afternoon with young Couchman by way of Liège. I went out to lunch with him and see him off. It is not an easy task he has ahead, but he went to it with a good heart.

Yesterday evening the Minister had an interview with Baron von der Lancken about the question of my making a statement as to what I saw at Louvain. I naturally am very reluctant to be brought into the affair, but the Germans have been very insistent, and finally von der Lancken said that he was confident that if he could talk with me for a few minutes he could arrange the matter to the satisfaction of everybody. He asked that I go to see him at the Ministry at half past six. I hurried home and dressed for dinner, so as to be able to go straight to Mrs. Z.'s, and then run over to the Ministry on the minute. The office of von der Lancken was dark and empty. I waited in the chilly corridors for twenty minutes and then went my way.

This morning one of his minions was here on another matter and I took occasion to mention the fact that he had not been there when I called. He came right back with the statement that they had come back from the field particularly early, on my account, and had waited for me in vain for nearly an hour. I assured them that I had been there on the minute and had been in the office, and that there was no one there. Mystery! By way of clinching it I said that the office was dark as the tomb. Then a ray of light struck the German, and he said: "Oh, I see, you came at half past six, Belgian time! Of course von der Lancken expected you at half past six, German time!!!" When he asked me when I would call I felt inclined to set eleven in the morning and then wander over at three in the afternoon, with the statement that, of course, I did everything according to New York time.

I had an hour's talk with von der Lancken about noon, and finally got off without testifying, which is a great comfort to me. He knew from their own troops that I had been in Louvain during the fighting, and had already reported that to Berlin. I finally prevailed upon him to let it go at that.

After we had settled our business, von der Lancken talked to me for half an hour or so about the war in general. He said they had just received a telegram that Reims is in flames, cathedral and all. It is a terrible thing to think of, and I suppose may turn out to be another Louvain before we get through. Von der Lancken explained it on the ground that French troops had come up and occupied the town, and that it was necessary to take it by storm—that troops could never operate against a position of that sort until artillery had cleared the way. I don't know just how far that sort of an explanation explains.

The Germans got out an *affiche* of news this morning, stating that *"les troupes Allemands ont fait des progrès sur certains points."* It does not sound very enthusiastic.

People coming in from Mons and Charleroi yesterday and to-day say that the German rear guard has fallen back on villages near those places and ordered the inhabitants to leave; the idea evidently being that they are preparing to resist any further advance of the allies.

After lunch, Baron de Menten de Horne was brought into the Legation again. The Germans seem anxious to get rid of him, and have finally turned him loose. I cannot very well make out their object in setting him free without getting a German officer in exchange, but they were keen to get him off their hands and wanted us to take cognisance of the fact that they had accorded him his liberty. This we have done. I shall be curious to see whether there is any sequel to this case.

Late this afternoon we got a telegram from the Consul at Liège, stating that Shaler and Couchman had been arrested in that city because they were carrying private letters to be posted when they got to England. They had taken a certain number of letters, all of them open and containing nothing but information as to the welfare of individuals here. They were on a mission of interest to the German authorities—getting foodstuffs to prevent a famine here. The Minister got off an urgent telegram to the Consul to get to work and have them released, and also saw von der Lancken about it, with the result that the wires are hot. I hope to hear to-night that they are free. These are parlous times to be travelling with correspondence.

I may have to get away any minute for Antwerp, to see if we cannot arrange to get flour down here for the city. There is enough for only a few days now, and there will be trouble when the bread gives out.

We have now been charged with Japanese interests; that makes six Legations we have to look after.

Wednesday.—Late yesterday afternoon I got a note from Princess P—— de B——, asking me to go to see her. I got away from my toil and troubles at seven, and went up to find out what was the matter. The old lady was in a terrible state. A member of her immediate family married the Duke of ——, a German who has always lived here a great deal. At the beginning of the war, things got so hot for any one with any German taint that they cleared out. For the last few days, German officers have been coming to the house in uniform asking to see the Princess. The servants have stood them off with the statement that she was out, but she cannot keep that up indefinitely. They are undoubtedly anxious to see her, in order to give her some messages from the ——'s, some of her other relatives in Germany; but if it gets around town that she is receiving officers in uniform the town will be up in arms, and the lady's life would be made miserable whenever the Germans do get out. She wanted me to start right away for Antwerp and take her along, so that she could send her intendant around afterward to say that she was away on a journey, and could not see the officers who had been sent to see her. I laboured with her, and convinced her that the best thing was to be absolutely frank. She is going to send her intendant around to see von der Lancken, and explain to him frankly the embarrassment to which she would be subjected by having to receive officers at her home. I am sure that Lancken will realise the difficult situation the old lady is in, and will find some way of calling his people off.

Went down to the Palace and had dinner with Pousette and Bulle and Cavalcanti, who were full of such news as there is floating around the town. There is a growing impression that the Germans do intend to invest Antwerp, and the Belgians are apparently getting ready for that contingency—by inundating a lot more of the country outside the ring of forts.

At noon, day before yesterday, I found a man with a copy of the *London Times*, and carried it in my overcoat pocket to the Palace Hotel when I went there to lunch. Last night, a lot of German civil officials were sitting at a table near by and holding forth in loud tones on the punishment that should be meted out to people who had forbidden newspapers in their possession. The most vehement one of the lot expressed great indignation that the *Amerikanischer Legationsrath* had been seen in that very restaurant the day before with an English newspaper in his overcoat pocket. Pretty good spy you have, Fritz.

A telegram has just been received from Liège, saying that Shaler and Couchman have been released and are on their way to Holland. A Dutch messenger was in after lunch, and told me that he had seen the two men at headquarters yesterday afternoon, and that they were far from happy. He said he did not blame them, as the Germans are dealing out summary justice to anybody who falls into their hands that they do not take a fancy to.

A. B. has been after me for a couple of days to take her up to the château near Louvain, where Countess R. is left alone with twenty-eight German officers quartered on her. A man cousin was sent up to defend her, but was so badly frightened that he spent all his time in the cellar and finally ran away and came back to Brussels. Now she wants to go up to the rescue, and stay there. I have asked von der Lancken for a pass, and shall try to take her up to-morrow. She certainly has good nerve, but I am not sure how much protection she would be able to afford.

The supply of flour is getting pretty well used up, and I may have to clear out to-morrow afternoon or the next day to go to Antwerp and negotiate to have some supplies sent down for the relief of the civil population. The Government has volunteered to do this, if the Germans would promise that the food would not be requisitioned for the troops. We have been given these assurances, and it only remains for me to go up and complete the arrangements.

When the Minister came back from Louvain he went over to headquarters and talked about the subject of my trip to Antwerp. He has been nervous about each of my trips and has worried a lot more about it than I have, but when he saw von der Lancken, that worthy made things worse by saying that there was artillery ready to begin business in every part of the country I was to traverse and that it would be a very dangerous trip. Now, the Minister is making superhuman efforts to find some other way to get the letters and papers through to Antwerp.

A note has just come in from Princess P. de Z——, to say that she followed my advice, and that everything has been settled with the German authorities to her complete satisfaction. She is now easy in her mind.

September 25th.—I spent all day yesterday sitting on the edge of my chair waiting for a decision about my leaving for Antwerp, and by dark I was a fit candidate for an asylum. At five o'clock the Minister went around to see von der Lancken to get the *laisser-passer*. It was then suggested that a letter could be sent around by way of Berlin and The Hague. It would take a week or ten days to get an answer that way. Then we argued the matter out again from the beginning, and after a quarter of an hour of joint debate I went over to see von der Lancken and press for the *laisser-passer*. He was in a *conseil de guerre*, but I had him pulled out and put it up to him. He said it was then too late to get anything last night, but that he would attend to it to-day. I am now sitting on the same old edge of my chair waiting for action, so that I can get away. I think that the trip by Namur, Liège and Maestricht, which is the route prescribed, is a lot safer than the other two trips I have made to Antwerp, which really were risky performances. Most of this trip will be in peaceful Holland and I do not contemplate any sort of trouble along the way.

By way of being ready I got passes from the Dutch Legation and the Burgomaster yesterday afternoon, and now all I have to do is take the German *Passier-schein* in my hand and start.

Yesterday evening I dined at the M.'s. Just the two of them and their daughter, who is married to a French officer. As is the case everywhere else, they talk nothing but war, and are most rabid. They have a daughter in Germany, but she does not seem to enter into their calculations, and all their thoughts are for France and Belgium. Their son, who is in the Belgian cavalry, has just got his corporal's stripes for gallantry in action. The old gentleman is bursting with pride. During the evening another old chap came in with a letter from his son, who is in young M.'s regiment; he had some very nice things to say about the young man's behaviour, and there was a great popular rejoicing.

The *London Times* came in during the evening, and there was a great revamping of war maps to correspond with the latest movement of troops. The daughter keeps the maps up to date, and does it very well, having picked up some training from her husband. She has different coloured lines for each day's progress and it is easy to see at a glance just how the positions compare for any given times.

This morning the Germans have big placards up all over town, trying to explain their action in burning Reims Cathedral. They are doing a lot of explaining these days.

Brussels, September 26, 1914.—My departure for Antwerp has been put off again and again, but if the German authorities live up to their promises, I shall be able to start to-morrow morning early. At the last minute the mothers of Mr. and Mrs. Whitlock decided to avail of the opportunity to go home, so I shall take them as far as Rotterdam before going to Antwerp. I shall attend to my business there and then go back to Rotterdam, take the ladies over to England, turn them over to Mr. N———, spend a day or two there getting a line on the news, and then rush back to Antwerp, and then back to Brussels. I suppose I shall be away ten days or so, but there is no way of telling. I should like the little trip to England and a breath of air in a country where there is no actual fighting.

It is now half past eight and there is no telling when this family will sit down to dine. The Burgomaster has indulged in some more repartee with the German authorities, and they, with their usual *finesse*, have put 'him in prison. Yesterday the Germans got out a proclamation announcing that since the city of Brussels had not settled "voluntarily," the whole of the forced loan imposed upon her no more requisitions should be paid in cash, as had been promised.* Max thereupon sat down and wrote a letter to the banks, saying that they were to pay nothing on the forced loan unless and until the Germans conformed to their part of the agreement. He further annoyed the Germans by putting up an *affiche*, giving the lie to a proclamation of the Governor of Liège:

The German Governor of the town of Liège, Lieutenant-General von Kolewe, caused the following notice to be posted yesterday:

"*To the inhabitants of the town of Liège.*

"The Burgomaster of Brussels has informed the German Commander that the French Government has declared to the Belgian Government the impossibility of giving them any offensive assistance whatever, as they themselves are forced to adopt the defensive."

I absolutely deny this assertion.

ADOLPHE MAX,
Burgomaster.

Lüttwitz replied to this by having Max arrested, and the present prospect is that he is to be sent to Germany as a prisoner of war. That is not very com-

---

* The German point of view was set forth in the following official notice:

"The German Government had ordered the cash payment of requisition, naturally believing that the city would voluntarily pay the whole of the forced payment (*contribution de guerre*) imposed upon it.

"It was only this condition that could justify the favoured treatment enjoyed by Brussels, as distinguished from the other cities of Belgium which will not have their requisition orders settled until after the conclusion of peace.

"Inasmuch as the city administration of Brussels refuses to settle the remainder of the forced payment, from this day forward no requisition will be settled in cash by the Government treasury.

Brussels, September 24, 1914.

"The Military Governor,
BARON VON LÜTTWITZ,
Major-General"

forting for us, as he has been a very calming influence, and has kept the population of Brussels well in hand. If they do send him away, the Germans will do a very stupid thing from their own point of view, and will make Max a popular hero everywhere.

Early this evening Monsieur Lemonnier, the Senior Alderman, came around with several of his colleagues, and laid the matter before Mr. Whitlock and the Spanish Minister. They immediately went over to see General von Lüttwitz to see whether there was anything to be done for Max, but as they have been gone a long time, I fear they are going through one of those long and thoroughly unsatisfactory discussions that get nowhere.

Monsieur Lemonnier is waiting in my office to hear the result of the visit to Lüttwitz. He is naturally far from cheerful, and looks forward with a good deal of dread to taking over the reins if Max is sent to Germany. He, of course, foresees that the chances are in favour of his following Max into exile sooner or later, if he tries to do his duty. As to his own future he says only—"I succeed only to the troubles of the office—*Max a bien emporté sa gloire avec lui.*" The life of a Belgian official these days is anything but comfortable.

Sunday Morning.—We were all up working until two o'clock this morning. Monsieur Max was spirited away to Namur, and everybody is standing by for trouble. The people are greatly excited and highly resentful, but it is to be hoped that they will not do anything rash. The cooler spirits are going about urging calm. The excitement is not lessened by the fact that there is heavy cannonading from the direction of Antwerp.

Lüttwitz has announced the arrest of Max in the following poster:

NOTICE.

Burgomaster Max having failed to fulfil the engagements entered into with the German Government, I am forced to suspend him from his position.

Monsieur Max will be held in honourable detention in a fortress.

> The Military Governor,
> BARON VON LÜTTWITZ,
> *General.*

Brussels, September 26, 1914.

We are evidently not yet through the epoch of destruction, for the Governor-General came out to-day with this Proclamation, which is posted on the walls of various towns:

Recently, in regions not occupied by strong forces of German troops, convoys of transport wagons and patrols have been attacked without warning by the inhabitants.

I draw the attention of the public to the fact that a list is kept of the towns and communes in the vicinity of which these attacks have been committed, and that they must expect their punishment as soon as German troops pass near them.

I have not been able to learn of any places where such attacks have taken place, but suppose this is merely an evidence of the well-known nervousness of the army of occupation, and that they are trying to frighten the people to a point where they will not try to start anything.

Fire at Namur during the bombardment

Effect of big German shell on Fort of Waehlem

Outside view of the Fort of Waehlem after bombardment by big German guns

General von Lüttwitz has come out with another Proclamation, forbidding the sale of foreign newspapers in Belgium:

I remind the population of Brussels and its suburbs that it is strictly forbidden to sell or distribute newspapers that are not expressly authorised by the German Military Government. Any infraction of this prohibition will entail the immediate arrest of the vendors, as well as long periods of imprisonment.

The German Military Governor,
BARON VON LÜTTWITZ,
General.

My *laisser-passer* has not come, and there is no telling when we shall get away. The Germans swear it was sent last night.

On board S. S. "Oranje Nassau," off Flushing, Sept. 30, 1914.—We got away on Sunday morning about eleven o'clock, after many calls at headquarters and a mild row about the *laisser-passer* that had not been sent. It was finally discovered that some boneheaded clerk had sent it by mail—a matter of three days! It was fished out of the military post office, and we got away in a few minutes.

We were in the big car, heavily laden—two trunks, several valises and a mail pouch on top—my two passengers inside with their small stuff, the chauffeur and I in front.

We made quick time out through Tervueren and down to Namur, hearing the heavy booming of cannon all the time away to the north. Ruin was all the way— odd farm-houses burned, towns with half the buildings in them, the Grand Place destroyed, etc. The great square at Namur a heap of brick and mortar.

The great bridge across the Meuse was dynamited, and the three sections hung in the river. All the way to Liège the main bridges had been destroyed, and we had to cross on temporary affairs constructed by the Germans.

And the Germans were thick all the way, holding us up at frequent intervals to look at our papers. They have it in for Belgium, and are in bad humour. We had some fine samples of it during the day.

We stopped not far from Huy for a picnic lunch, and then got under way again, being stopped frequently all the way to Liège, where we sought out the Consulate. The Consul had gone to Spa to look after some English people, but I said my few words to his wife and daughter, and then hurried away toward Visé and the Dutch frontier.

Visé n'existe plus! Goodness knows what was done to the place, but there is nothing left but blackened walls. It took us a long time to find unencumbered roads and get through between the fallen walls. Not far from the edge of town we found the last German outpost, and were promptly put under arrest because my *laisser-passer* did not bear my photograph. The officer in command cursed me roundly for daring to come through Liège without reporting, placed two armed soldiers in the car, and ordered us sent back. It was futile to point out to him that passes issued by the Military Governor General did not need to conform to the local rules; in fact, it only made him peevish. We scorched back over the road to Liège, but I succeeded in making the soldiers stop at a small town where there was a local headquarters of some sort with a colonel in command. I got him to look at our pass which had been confiscated by our guard, and, after hearing my case and thinking heavily, he unenthusiastically said we might proceed. We went back through Visé even faster, and enjoyed the look of our lieutenant when told he had been overruled. After a minute or so he became very affable and said he had a brother in Jefferson City, Mo., and a nephew in Sacramento, *Californien*, who runs an *Apoteke*. Just to show there was no hard feeling, I gave him a cigar, and a few minutes later we crossed the Dutch frontier, where we created a sensation. A big crowd gathered around the car, and, by the time the leisurely custom officers had examined the papers given me by the Dutch Legation, they were packed so tight that it took the united effort of several officers and citizens to get us extricated.

Holland is taking no chances, and has quantities of troops massed in that part of the country. There are frequent posts to stop travellers and examine papers, and there is practically no traffic on the road save that of a military character.

Near Maestricht we ran into a large detachment guarding a bridge. Our papers did not satisfy the commanding officer, so we were once more placed under arrest and hustled through town to headquarters. The officers there were very courteous, and, after examining my papers, made out a *laisser-passer* for use in Holland and sent me on my way.

By this time it was dark, but we determined to push on as far as Roermond—50 kilometers. Here we found a charming little hotel—the Lion d'Or—and after a good supper, got early to bed.

The next day I planned to take the two ladies—who have good nerve, and don't turn a hair at being arrested—to Rotterdam and then run down to Antwerp, some 280 kilometers, a long run in war time.

We were off at 6:30, and bowled along beautifully in a bitter cold wind until we were in sight of Tilburg, where the engine broke down. Eugène, the chauffeur, tried everything he could think of, and tore his hair in rage and shame. Finally we got a soldier on a bicycle to go into Tilburg and get a motor to tow us in. Then two good hours in a garage before we were in shape to start.

We caught the boat at Moerdyck and got into Rotterdam a little before four. I installed my companions at the Maas Hotel, overlooking the same old Meuse, and then started back through the rain toward Antwerp. At Willemsdorp we just missed the boat for Moerdyck and lost an hour. Eugène raged and smoked many cigarettes, to the danger of his health, because his *sacrée* machine had lost us so much time.

At eight we got to Rosendaal, near the Belgian frontier, and were forbidden to go any farther until morning, as the outposts were taking no chances.

Had a good supper at the little hotel, had my papers viséed by the Belgian Consul, and at 6 o'clock yesterday morning was up and away, by way of Putte.

The Belgian outposts received us with levelled rifles, but when we got near, one of the officers recognised me through his glasses, and we got through without any more trouble. Arrived at the St. Antoine as everybody was coming down to breakfast. The Germans were bombarding the outer forts, and they could not believe their eyes when I came in. Not a word of news had got through the lines for some days, and I was nearly torn to pieces by the excited friends.

I had coffee with Colonel Fairholme, and got all the news he could tell me. Malines has been bombarded again, and Antwerp is filled with refugees. Before I left, the Germans had occupied Malines itself and were bombarding the fort at Waelhem.

After breakfast I started out on my carefully planned campaign. First to the Consulate-General to get off some telegrams, etc. Then to the Foreign Office with a lot of things to attend to. I was able to give van der Elst word that his son is in Magdebourg—a prisoner, but not wounded. The look on his face when he got the news paid for the whole trip. I saw M. Davignon, and went with him to see the Prime Minister, who had heard I was there and had sent for me.

On the way we saw hundreds of miserable refugees from Malines pouring down from the station. The courage of these Belgians is beyond all words. Save for the two in the freight station yard at Louvain, I have not seen a woman crying! It may be that they are numb, but they have none of the stupidity of numbness. And when you think that these very women will be creeping back to their homes and caring for the German wounded they find there, it gives you a fine lump in the throat.

I paid a call at the French Legation, went back to the Consulate-General to sign my telegrams and mail which had been hammered out, and then to lunch. Got away at 3:30 to the banging of heavy siege artillery and invitations to come back "if we are still here." As I was getting into the car, Prince D—— plucked me by the sleeve and pointed at the Cathedral tower high above us. "Take a good look," he said. "It may not be here when you come back."

We made good time through the rain, but missed the boat at Moerdyck, and spent an hour on the dock. Got in at ten, ravenously hungry, had a snack, and then to bed.

Up again at six and took the seven-thirty train for Flushing. It loafed along through the country, and we did not sail until eleven. We have to go round to Folkstone, but hope to be in by six o'clock.

There are not more than twenty people on the ship, and the way they went through our credentials was a caution. I was glad I had taken the precaution to provide myself with American, British, German, Dutch and Belgian papers for the trip. There is another examination at Folkstone.

On board the S.S. "Brussels," off Flushing, October 5, 1914.—To resume.

We got into Folkstone last Wednesday evening at sunset, and got through to London by eight-fifteen. All the latter part of the crossing we were spoken from time to time by British destroyers, which bobbed up from nowhere to warn of floating mines or give directions as to our course. The entrance to Dover was surrounded by destroyers, and looked grim and warlike, and what's more, businesslike.

Thursday morning I got up as late as I decently could and went down to the Embassy to find Shaler and Couchman waiting for me. They had been in London since Monday, but had not made much progress with their mission of getting food for Brussels. This was due to no lack of energy on their part, but to the general difficulty of getting attention for any matter at this time. I went with them to the Belgian Legation, and after a talk with the Belgian Minister, we got things started.

As the food was intended for the civil population of Brussels, it was necessary to get the Belgian Minister to secure from the Foreign Office permission to ship it through the blockade. He felt that he must have some instructions from the Government at Antwerp for his guidance in the matter, so I telegraphed at some length, with the result that he had ample instructions before the sun went down. The next day he made three or four calls at the Foreign Office and matters were got under way.

Shaler is buying the food and getting it ready for shipment, and now all that is holding things up is the actual permission to go ahead and ship. Shaler has had some talk on the general problems that confront us with Herbert Hoover, an American mining engineer, who has given some very helpful ideas and may do more still.

Shaler and Couchman had an experience at Liège they did not particularly relish. They were pulled up by a Landsturm guard somewhere in Liège, taken to the Kommandantur, where it was discovered that they were carrying a number of messages of the "We-are-well-and-hope-you-are-the-same" variety. Without discussion they were pushed into cells and treated to talk that gave them little comfort. They spent the night in jail, but by some means contrived to get word to the Consul, who arrived and delivered them before breakfast. It evidently grieved the Germans that they could not take these two out and shoot them, but they yielded with a bad grace and turned them loose to hasten to the Consul's breakfast table.

Brussels, October 11, 1914.—On Saturday afternoon late I went with Harold Fowler to call on Sir Claude MacDonald, who had been to the Embassy twice to see me about the English Red Cross nurses in Brussels. I tried to reassure him as to their safety, but he went to see the Ambassador later in the day and asked him to send Harold Fowler back to Brussels with me to bring the nurses out. This suited me perfectly, so we made preparations to get off together.

On Sunday evening we left Fenchurch Street at six, with a little group of friends to see us off. About the only other people on the train were a King's Messenger, a bankrupt Peer and his Man Friday, and a young staff officer. Each set of us had a separate compartment and travelled in lonely state to Tilbury, where the boat was waiting.

As we got aboard the *Brussels,* her sister ship, the *Dresden,* just in from Antwerp, pulled up alongside, and Mrs. Sherman, wife of the Vice-Consul, called me to the rail to give me the latest news. She said that everything was going to pieces, that some of the forts had fallen, and that Antwerp might be under bombardment before we got there. Then she went ashore in peace, and we went below to seek the seclusion that

View of the Meuse at Huy

Refugees fleeing toward Dunkirk before the German advance, after the fall of Antwerp

the cabin grants, and fortify ourselves for the bombardment.

We got under way during the night and dropped down to the mouth of the Thames, where we lay to until daylight, before starting across. The first sound I heard was a hail from a torpedo-boat destroyer, which sent an officer aboard to lay our course for us through the British mine fields. We made our zigzag course across the North Sea and fetched up at Flushing, where we picked up a pilot to take us through Dutch waters. When darkness overtook us we were just about on the Belgian frontier line and had to lie to for the night, getting to Antwerp Tuesday morning about nine.

We found the place in a great hubbub—everybody packed and ready to leave. They had been on the point of departure since Friday, and the uncertainty had got on everybody's nerves—and no wonder.

Several thousand British Marines had arrived and were doing good work, holding back the Germans, while the exhausted Belgians pulled themselves together for the evacuation. The Belgian forces had been fighting with little rest and no sleep until they were physically incapable of further resistance. How human strength held out so long is the great marvel. Winston Churchill was in the Legation when I arrived, with General Rawlinson and Colonel Seeley.

After a call at the Foreign Office, most of which had been installed on a boat in the river, I went to the Palace to see General Jungbluth. He was not there, but Countess de Caraman-Chimay said that the King wanted to see me.

I was taken straight up to him in his Council Chamber, where I found him seated at a great table covered with maps and papers. He pushed them aside wearily as I came in, and rose to greet me. He talked at some length on the war and the ordeal of Belgium, but was chiefly interested in how the people were being treated. His interest was not only for his own friends, but he showed particular interest in learning how the poorer people were being treated—whether the poorer quarters of the town were keeping calm and avoiding trouble with the Germans. He was most anxious that they should avoid doing anything that would arouse the Germans against them. He spoke simply and touchingly of his confidence in the loyalty and patriotism of all his people, and his certainty that they would come through the war with an even greater love of country.

The rest of the Palace was in confusion, with servants packing and orderlies coming and going. But the King's room was in perfect calm. The King sat quite still in his armchair and talked quietly, without haste. He was very serious, and it was clearly to be seen that he felt his responsibility and the suffering of his army. But his determination was just as evident. He realised that the evacuation was inevitable, and having made up his mind to that, he devoted his whole energies and thoughts to seeing that it was carried out effectively and quickly. He has a very patent faculty of concentration and of eliminating his own personality and feelings. I have seldom felt so sorry for anyone, partly perhaps because all of his sympathy was for others.

When the King finally rose to dismiss me, he said:

"The Queen wants to see you. Will you come back at half-past two?"

I had planned to leave for Brussels immediately after luncheon, but, of course, this was a command to which I gladly yielded.

The St. Antoine was all hurry and confusion, and the dining room was buzzing with conjecture as to whether the bombardment of the city would begin before the exodus was accomplished. The Military Governor had posted a proclamation to warn the population that it might begin at any time. There was a certain amount of unconscious humour in his proclamation. He advised people to retire into their cellars with bedding, food, water and other necessaries; to disconnect the water, gas and electricity; to stuff the staircases with mattresses, as a matter of protection; to take with them picks and shovels, so that they could dig themselves out in case their houses fell in; and after a few more hints of this sort, the Governor genially remarks:

"Having taken these precautions, the population can await the bombardment in calm."

The German authorities have offered to spare the historic monuments of Antwerp in their bombardment, if the Belgian General Staff will send them maps of the city with such monuments and hospitals clearly marked. I found that it had been arranged in Brussels that I should collect the plans on my way through Antwerp and deliver them to the German authorities in Brussels, and, of course, agreed to do so.

After luncheon I went back to the Palace, where I was immediately received by the Queen in her sitting room. Her Majesty seemed quite oblivious of the confusion in the Palace, and, like the King, she was chiefly concerned as to the welfare of the people left under German domination. I was able to give her comforting news as to the treatment of the people of Brussels. While we were talking, the roar of the German guns seemed to increase and made the windows rattle. There was an outcry in the street, and we went to the window to see a German aeroplane pursued by a British machine. We watched them out of sight, and then went back to our talk. The members of the Court had tried to prevail upon the Queen to leave Antwerp, but when it became evident that the place must be surrendered, she refused to move and told me she would stay until the King left. And she did.

When I got back to the hotel, I found Eugène with news that the differential of my car had broken, so that we could not start. It was important that we lose no time in getting the plans of the town to the German authorities, so I got Baron van der Elst to go with me to the General Staff and explain the situation. General de Guise promptly wrote out an order that I should be given the best car to be found in the city. Armed with this, Eugène set forth and gathered in a very pretty little limousine to bring us back to Brussels. It was evidently a lady's car and almost too pretty, but we were not exacting and took it thankfully. However, it was too late to start out through the lines, so we gave up the idea of leaving

Graves of civilians shot by the Germans

PROCLAMATION

A l'avenir les localités situées près de l'endroit où a eu lieu la destruction des chemins de fer et lignes télégraphiques seront punies sans pitié (il n'importe qu'elles soient coupables ou non de ces actes.) Dans ce but des otages ont été pris dans toutes les localités situées près des chemins de fer qui sont menacés de pareilles attaques; et au premier attentat à la destruction des lignes de chemins de fer, de lignes télégraphiques ou lignes téléphoniques, ils seront immédiatement fusillés.

Bruxelles, le 5 Octobre 1914

Le Gouverneur,
VON DER GOLTZ

A typical proclamation

Translation:
In future, villages in the vicinity of places where railway and telegraph lines are destroyed will be punished without pity (whether they are guilty or not of the acts in question). With this in view hostages have been taken in all villages near the railway line, which are threatened by such attacks. Upon the first attempt to destroy lines of railway, telegraph, or telephone, they will be immediately shot.

The Governor,
VON DER GOLTZ

Views of the Fort of Wahlem after its bombardment by the big
German guns

before morning. We had thought of taking the route of the army and getting to Brussels by way of Ghent, but the people at the General Staff said the road was so crowded with transport that we would make little progress, and that the better course would be to take exactly the opposite direction and go by way of Tournhout.

I took several of the ladies of the corps down to the boat, which was to take them to Ostend, which was to be the next stand of the Government. They all took it coolly and went to bed, as though there were no bombardment going on. The King and Queen, the Prime Minister, and the representatives of the allies remained in town overnight.

On one of my trips out of the hotel I met the Queen coming in to say good-bye to Princess Koudatcheff (wife of the Russian Minister), who was ill. She stopped to greet us and make inquiries as to each one.

After dark the crowd began to melt. Winston Churchill came down with his party, got into motors, and made off for Bruges. The Belgian officers staying at the hotel got off with their units, and by ten o'clock the staff of the British Legation, Fowler and I, were left in almost undisputed possession of the hotel. The water-supply was cut. The lights were out and the place was far from gay, particularly as nearly all the servants had fled, and we could not get anything to eat or drink.

Most of the town repaired to the cellars for the night, but we decided that if it really came, we saw no choice between going down with the house into the cellar and having the house come down on top of us, so we turned in and got a night's rest, which, I am free to confess, was rather fitful.

All night long motors were snorting away, and all night long the guns kept pounding, although they did not seem to get any nearer. With the intelligence that one has when half awake, I carefully arranged a pillow between me and the window, as a protection against shells!

We got up early and went out into the streets to watch the movement. The few remaining troops were being poured out on the road to Ghent. On foot, in motors, on trains, on bicycles, and on horseback, they streamed. The civil population was also getting away, and all the trams in the direction of the Dutch frontier were loaded with people carrying their little bundles— all they could hope to take away with them. The hospitals were being emptied of the wounded and they were getting away as best they could, those whose legs were all right helping those who had trouble in walking. It was a depressing sight, and above all, the sound of the big guns which we had heard steadily since the morning before.

We got under way about half-past eight, after a wretched and sketchy breakfast, and after saying good-bye to one of our friends of the British Legation.

First, we went to the north gate, only to find that it had been closed to vehicles a few minutes before, and that barbed-wire entanglements had been stretched across the road. Argument was vain, so we worked our way back through the traffic and reached the Porte de Tournhout, only to be turned back again. For nearly an hour we wandered about in the stream of refugees, in vehicles and on foot, before we finally succeeded in making our way through a side door of the Porte de Tournhout, and starting that way. We were not at all sure that we should be able to reach the Dutch frontier through Tournhout, as the Germans were supposed to be that far north, but we did make it after a long series of stops, to be examined by all sorts of Belgian outposts who kept cropping up out of fields to stop us and look through our papers. From some little distance out of town, we could see the shells bursting over the southern part of the town, or possibly over the villages to the south of the town proper.

We plowed along through Holland, being stopped all afternoon by Civil Guards, and reached Maestricht at sunset. We went straight to the German Consulate to have our papers put in order and learn whether it could be arranged for us to pass the lines at night. Our papers were not in order because they bore no photographs, and the Consul could not see that the German interest in our mission made any difference, so that there was nothing to do but wait over until morning, and get some pictures.

It took us until ten in the morning to get our photographs and have our papers arranged, and by good driving we reached Liège in time to lunch with the Consul. Then on to Brussels by way of Namur. On the road we picked up a German officer on his way to Namur, which kindly deed saved us much delay in being stopped by posts.

We reached Brussels at five and hastened to send the precious plans of Antwerp to Lancken. We had just settled down at the Legation to a good talk when word came that Lancken was anxious to see me at once. I went over to the Political Department to find that the gentleman merely wanted a formal statement from me as to when I had received and delivered the plans, so that he could make it a matter of record. I satisfied him on these points and went my way.

Then we gathered at the Legation and talked steadily until after midnight.

While I was away the Minister had got off a trainload of Americans, and with them he had sent the English nurses. That relieved Harold Fowler of the mission that brought him, but we bore up bravely.

The Germans have announced the fall of Antwerp and have apparently occupied the city. At first everybody was much downcast, but on second thought they have been convinced that the evacuation of the army and the surrender of an empty shell was a pretty clever piece of work. With the big siege guns that were in action, it was only a question of days until the Germans would have reduced all the forts. And then if the resistance had been maintained, the greater part of the army would probably have been captured. As it is, the Belgians inundated the country to keep the Germans from cutting off their retreat, and made off for Ostend, leaving only a handful of men with the British Marines, to hold the Germans in check. So far as we can learn, most of the army has succeeded in getting away and forming a junction with the allies.

Brussels, October 14, 1914.—We are quite up in the air about what we are to do next. Monday afternoon I went around to headquarters to get a *laisser-passer* to take Harold Fowler back to England. While the matter was being attended to, an officer came in and told me that Baron von der Lancken wanted very much to see me. When I went into his room, he said that there was nothing in particular that he wanted to see me about, but that he thought I would be interested in hearing the news and in telling him something of my trip. We talked along for some time about things in general, and then he told me that the movement of troops toward the coast was progressing rapidly and that the Belgian Government would soon be driven from the country. Then putting the tips of his fingers together and looking me coyly in the eye, he inquired: "And then my dear colleague, what will be your position?" He elaborated by pointing out that the Government, to which we are accredited, having left the country, we would be merely in the position of foreigners of distinction residing here, and that we would have no official rank or standing. The idea evidently is that they do not care to have us around any longer than they can help.

I later learned that Villalobar had been more ready than I with his retort. In the course of a call later in the afternoon, Lancken had talked the same matter over with him, and had wound up with the same genial question: "And then my dear colleague, what will be your position?" Without any hesitation, Villalobar replied: "My situation will be just the same as yours. We are both representatives of our country in a country not our own. We shall continue to owe each other respect, and to make the best of conditions."

The latest news we have this afternoon is to the effect that the Government has been driven from Ostend, presumably to the Isle of Guernsey. It would be pleasant, in a way, to retire to a retreat of that sort for a few months' rest, but I fear there is nothing of that sort in store.

To-day I ran across an order from the Governor-General forbidding civilians to ride bicycles. The order concludes as follows:

> Civilians who, in spite of this, continue to ride bicycles, expose themselves to being shot by German troops.
>
> If a cyclist is suspected of planning to damage railroad, telegraph or telephone lines, or of the intention of attacking German troops, he will be shot according to martial law.

Apparently it is no longer necessary to go through the forms of proving that the cyclist had any evil intention. The mere suspicion is enough to have him shot.

In the course of a visit to General von Lüttwitz to-day, one of the colleagues remarked that the Germans *must* keep the Belgians alive, and could not allow them to starve. Lüttwitz was not at all of that mind, for he said with some show of feeling:

"The allies are at liberty to feed the Belgians. If they don't, they are responsible for anything that may happen. If there are bread riots, the natural thing would be for us to drive the whole civil population into some restricted area, like the Province of Luxembourg, build a barbed wire fence around them, and leave them to starve in accordance with the policy of their allies."

And as the German policy is more or less frankly stated as a determination to wipe out as many of the enemy as possible without regard to what is or has been considered as permissible, it is quite within the realm of possibility that they would be prepared to let the Belgian people starve. In any event, you can't gamble with the lives of seven millions of people when all you have to go on is the belief that Germany will be guided by the dictates of humanity.

Fowler was to have left yesterday morning, and had engaged a seat in a new motor that is being run out by way of Maestricht. It was to have called at my house at seven o'clock yesterday morning, and we were up and about bright and early. We waited until a little after nine, when Eugène turned up to say that the chauffeur had been arrested and put in jail for having carried correspondence and having been caught nosing around one of the forts at Liège. The service is now suspended, and we don't see any prospect of his getting off before Friday, when we are sending a courier to the Legation at The Hague.

Yesterday afternoon we went up to Antwerp to see how our old motor-car was getting along. It was out of whack, and we were obliged to get another to come back to Brussels. I took the big car and organised an expedition of Monsieur de Leval, Fowler and a German official named Conrad, who went along to help us over the rough places. It is the first time for weeks that the direct route has been feasible.

I have had enough of ruined towns, and was not able to get the awful sights out of my head all night, but spent my time in bad dreams. From Vilvorde right into Antwerp there is not a town intact. Eppeghem, Sempst, Malines Waehlem, Berchem—all razed to the ground. In Malines a good part of the town is standing and I suppose that the Cathedral can be restored, but the other towns are done for. There were practically no civilians in any of them—a few poor peasants poking dismally about in the ruins, trying to find some odds and ends that they could save from the general wreck. There were some children sitting on the steps of deserted houses and a few hungry dogs prowling around, but no other signs of life. All the way from the outskirts of Brussels straight through to Antwerp, the road was lined with empty bottles. They gave a pretty good idea of what had gone on along the line of march.

The bombardment of Antwerp lasted from the afternoon that we left up to Friday noon. The damage is pretty evenly distributed. Houses here and there in every street were badly smashed and the whole block across the street from the Hôtel St. Antoine, where we stayed, was burned to the ground. The Cathedral was not damaged.

When we were there last week, the streets were thronged with people and with motors. Yesterday there was not a soul to be seen for blocks together. The town was practically deserted.

The garage where I had left my car had been taken over by the military authorities. The car was put away on the second floor undamaged, but also unrepaired, so we shall have to wait until things settle down a little and we can get some work done. I shall have to go back to Antwerp a little later and attend to that. There is some comfort in the fact that the car has not been smashed.

This morning the Committee for the Provisioning of Brussels came in, and asked whether I was prepared to go to London for them and endeavour to arrange for some sort of permanent agreement with the British Government for the provisioning of the civilian population of Belgium. I am willing.

In the course of some errands this afternoon, I dropped in on Baronne Lambert for a cup of tea. The Baron came in and then Villalobar. About two minutes later, Lambert was called out of the room to speak with a German officer, who demanded that he accompany him to headquarters. Villalobar went with him to see what was up, and I stayed behind to see if I could be of any use. We stood by for a little over half an hour, and then when Mme. Lambert could stand it no longer, I jumped in my car and went down to see what was happening. I found Villalobar on the sidewalk, getting into his car. He was depressed and said that he had been obliged to leave the Baron with the Germans; that he was suspected of nobody would say what, and that the Germans were going to search the house. I went back and had them all ready for the shock of the invasion. They were standing by for the search party, when in walked the Baron, smiling broadly. They had sent him home under guard of two armed men, and were to search the house in the course of a few minutes. While he was telling about it, two officers arrived, profusely apologetic, and asked to be shown over the Red Cross hospital, which had been installed on the ground floor. They were taken all through the place, and found only a lot of German soldiers carrying off the beds and other belongings. Then they searched the Baron's private office and that of his son, and withdrew after more excuses.

There was nothing to show for the whole performance, and nothing had been accomplished beyond making a lot of people nervous and apprehensive. That is the sort of thing that everybody is subject to these days, without any hope of redress. And, of course, this was the least serious thing that could happen.

On board S.S. "Princess Juliana," off Dover, Sunday, October 19, 1914.—Here we are again, coming into England in rain and fog. Up to the last minute, I was in great doubt as to whether we should come at all, but everything was finally straightened out and here we are.

Friday we spent in hard work, aggravated with many conferences. In the morning most of the German civil and military Government came to the Legation and discussed the food question with the members of the Committee, the Spanish Minister and ourselves. They all united in asking that I go to London and lay the situation before the Belgian Minister, the Spanish and American Ambassadors and, under their chaperonage, before the British Government. When this had been agreed to, some bright soul suggested that I be accompanied by a commission of fifteen prominent Belgians, to add impressiveness to what I had to say. The two Ministers rose up and said *no*, adding that as I was to do the work and bear the responsibility in going on this mission of forlorn hope, I should not be hampered by having to carry the weight of fifteen speech makers. That was knocked in the head, and then to show that we were not unreasonable, we asked that two members of the Committee go along. The men chosen were Baron Lambert and Monsieur Francqui, one of the leading bankers of Brussels and a man of poise and judgment. They expressed reluctance but were soon persuaded.

This morning, during a call at the Political Department, the talk turned on Mexico. I was asked what the President was driving at, and answered that he was clearly trying to give the Mexicans every opportunity to solve their own troubles without interference. I was then asked, rather slyly, whether the President really wanted them to settle their troubles. Without waiting to hear my answer, the oracle went on to tell me what our real policy was as he saw it, and he had no doubts. The President wanted to take Mexico, but was intelligent enough to realise that if he simply seized it, he would forfeit any claim he might have to disinterestedness, and our Anglo-Saxon hypocrisy could not swallow that. Therefore, he was deliberately allowing the Mexicans to drift into a hopeless condition of anarchy, which he knew would get steadily worse, until all the best and most prosperous elements in the country would come to the conclusion that they would be happier and safer under American rule than under the uncertain despotism of changing factions. The President could then yield to their entreaties, and could take over the government of Mexico as a humanitarian service to the people.

I made a feeble attempt to explain what our real feelings were toward Mexico, but it soon became evident that we could not think in the same terms, so I gave up. There was no criticism expressed or implied. On the contrary, there was evidence of real admiration of the President's technique.

The rest of the day was spent in getting ready letters and telegrams and other papers necessary in our work.

Fowler and I dined at the Lambert's, finished up our work at the Legation, and got to bed at midnight. We got up yesterday morning at half-past three, and at half-past four set sail in three motors—one filled with servants and mountains of small baggage.

We sped in the dark through ruined villages to Antwerp, and from there to Esschen on the Dutch frontier, which we reached soon after daylight. We had papers from the Dutch Legation, calling upon the customs authorities to let us pass, but a chuckle-headed *douanier* would not even read our papers, and held us up for an hour, while he made out papers of various sorts and collected a deposit on our cars. I expostulated in vain, and shall have to get my comfort from making a row later. As a consequence of his cussedness, we missed the morning boat train to Flushing, and had to spend the day in that charming city. We found the place filled with refugees from all parts of Belgium, and were greeted on every hand by people we knew. The hotels were filled to overflowing, and people were living in freight cars, sheds and on the sidewalk. We clung to chairs in the reading room at one of the hotels, and walked the streets until nine o'clock, when we got aboard the boat with eight hundred other people. Cabins were not to be had for love or money, but Francqui, by judicious corruption, got us a place to sleep, and we slept hard, despite the noise, which was tremendous.

London, October 20, 1914.—Here we are, much cheered up by the prospect.

We hammered hard yesterday and to-day, and this afternoon it looks as though we had secured the permission of the British Government to send food to our people in Belgium.

We got into Folkstone at 4 o'clock on Sunday, were passed immediately by the authorities, and then spent an hour and a half waiting for our train to pull out. We got into darkened London about a quarter of eight. We sat around and visited beyond our usual hours, and yesterday morning I was called ahead of anybody else, so as to get down to my day's work.

First, I got things started at the Embassy, by getting off a lot of telegrams and running away from an office full of people who, in some mysterious way, had heard I was here. I saw several of them, but as my day was going, I up and ran.

First, to Alfred Rothschild's house in Park Lane, where I found Baron Lambert waiting for me. He was beaming, as his son (serving in the Belgian army) had turned up safe and well before leaving to rejoin his regiment in France.

Next I went to the Spanish Embassy, and gave the Ambassador details of what we wanted. He caught the idea immediately, and has done everything in his power.

When I got back to our chancery, I found that the Ambassador had come in, so I went over the whole business again, and made an appointment for a conference with him for the Spanish Ambassador and my travelling companions

At half-past five we had our conference with the two Ambassadors. They made an appointment with Sir Edward Grey for this afternoon, and went over the situation at some length, to make sure of the details.

In view of its significance this meeting was most impressive to me. It was made up of the two Ambassadors, my two companions, and Herbert Hoover, the man who is going to tackle one of the biggest jobs of the time. He has been studying the situation, the needs of the civil population and the difficulties to be overcome ever since Shaler's arrival several weeks ago. While we could enlighten him in regard to recent developments and matters of detail I was astonished to see how clearly he grasped all the essentials of the situation. He sat still while the rest of us talked but his few remarks were very much to the point, particularly when, in answer to a question, he said very quietly: "Yes, I'll take over the work. I have about finished what I have in hand. Now we can take up this."

October 21st.—The Belgian Government has sent over Monsieur de Berryer, the Minister of the Interior, to discuss the food question and the equally important money question.

I had an early morning note from the Spanish Ambassador and went around to see him.

London is filled with war spirit; not hysterics, but good determined work. The streets are full of singing recruits marching hither and yon—mostly yon. The army must be growing at a tremendous rate; in fact, faster than equipment can be provided, and they are not slow about that.

London, October 23, 1914.—On Wednesday we had things pretty well settled, and had also succeeded in raising from official sources about one hundred and fifty thousand pounds. I took a fair amount of satisfaction in gloating over those who had croaked. Then some helpful soul came along and threw a monkey wrench into the machinery, so that a good part of the work has to be done over again. At any rate, we hope to get, some time to-day, permission to export enough food to serve as a stop gap until the general question can be settled.

Monsieur Francqui and Baron Lambert had to start back this morning to organise the Belgian local committees into one central national affair, and I am to stay on until things are settled one way or the other. That may mean not getting back to Belgium for a week or two more.

For some time I have been threatening to get a dog and yesterday, feeling the need of intelligent canine sympathy, I succumbed. At the Army and Navy Stores, I found a hideous brindle bull that some officer had left on going to the front. He was promptly acquired, and given the name of Max in honour of our Burgomaster. The Stores are to take care of him for me until I return to Belgium.

When I got back to the Embassy, from my visit to the Stores, I found Shaler waiting for me with the news that I was expected at a meeting at Mr. Hoover's office in fifteen minutes, to discuss matters with the committee which is being formed to handle the feeding of the Belgian civil population.

I was surprised to find that I had been made a member of this committee, and was expected to attend. It was a comfort to talk with men who know what they are about and who can make up their minds right the first time. Hoover is a wonder and has the faculty of getting big-calibre men about him. We were not in session more than an hour, but in that time we went over the needs of the Belgian civil population, the means of meeting immediate needs, the broader question of finding food from other parts of the world to continue the work, the problem of getting money from public and private sources to pay expenses, and finally the organisation to be set up in Belgium, England, America and Holland, to handle the work. Before we left a tentative organisation had been established and people despatched on various duties with orders to get things started without loss of time, so that food could be pushed across the line into Belgium at the first possible moment.

It is going to be up-hill work for many reasons, but it would be hard to find a group of men who inspire as much confidence as these that everything possible will be done, and occasionally a little that is impossible.

October 24th.—Yesterday was another busy day. I did not know that the entire population of Belgium

Herbert C. Hoover

French Howitzer near H——

German camp kitchen

could make such a crowd as I have had in the waiting-room of the chancery. In some mysterious way the news of my coming to London has got about, and swarms of people are coming in with little errands they want done and messages to be delivered to their friends and families in Brussels. It makes work, but that sort of thing is a comfort to lots of people and is worth undertaking. I have made it clear to all of them that anything to be delivered will be turned over to the German authorities first, and hope they will govern themselves accordingly.

The British Government has stipulated that the feeding of the civil population shall be carried on by a neutral organisation, under the patronage of the American and Spanish Ambassadors in London and Berlin, and the American and Spanish Ministers in Brussels. The food is to be consigned to the American Minister in Brussels for distribution by the organisation which is to be known as the American Relief Committee, with Hoover as chairman and motive power. The various local Belgian committees are to be grouped together in a national organisation, to assist in the distribution of the foodstuffs once they are delivered inside the Belgian frontier. The members of the Belgian organisation are, of course, prisoners of the Germans and unable to give any effective guarantees as to the disposal of the supplies. The British Government has, therefore, stipulated that all authority and responsibility are to be vested in the American Committee, and that the Belgians are to be regarded simply as a distributing agency. This is, of course, in no sense a reflection of the Belgians engaged on the

work, but merely a recognition of the difficulties of
their position.

The neutral composition of the Committee assures
it a freedom of travel and action, and an independence
of political and personal pressure, and a consequent
freedom of administration which the Belgians could
not hope to enjoy. It is only by the assumption of
complete authority and responsibility by the Committee that the patrons will be able to give the various
Governments concerned the necessary assurances as
to the disposition of foodstuffs and the fulfillment of
guarantees.

There is something splendid about the way Hoover
and his associates have abandoned their own affairs
and all thought of themselves in order to turn their
entire attention to feeding the Belgians. They have
absolutely cut loose from their business, and are to
give their whole time to the work of the Committee.
This is done without heroics. I should hardly have
known it was done, but for the fact that Hoover remarked in a matter of fact way:

"Of course everybody will have to be prepared to
let business go and give their whole time."

And it was so completely taken for granted that
there is nothing but a murmur of assent.

Another strenuous day on the food question and
other things.

My plans were to leave for Brussels on Monday
morning, but in the evening the Ambassador sent
for me and it was decided that I should go to Havre
and from there to see the King and Queen. That
will take me to within a couple of hours from Brussels,

according to old calculations, but under present conditions I shall have to get there by way of France, England and Holland.

Hôtel des Régates, Havre, October 26, 1914.—This is the third town where I have paid my respects to the Belgian Government. I would gladly have foregone the experience, for it is depressing.

I left Waterloo station at 9:15 last night. Instead of the usual two-hour run to Southampton, we puttered along and did not arrive until after one. I had a compartment and made myself as comfortable as possible. When we arrived I found poor Colonel Swalm, the Consul, waiting for me. The Ambassador had telegraphed him to see me off, and he did so regardless of the hour. I felt horribly guilty to have him waiting about for me, but it certainly did make things a lot easier.

I got straight to bed, but had a hard time sleeping, as there was a tremendous racket of loading all night long. Nearly all the passengers were British officers on their way to the front. Among the others I found de Bassompierre of the Foreign Office, and a Mr. and Mrs. W——, who were coming over with a Rolls-Royce, to be presented to the Belgian General Staff. If I go to the front, he will take me. We sailed at daybreak and were here by two o'clock. Our Consul, Osborne, was waiting for me at the dock with Henry Needham, the correspondent of *Colliers*. I was let straight through the customs, where a *woman* marked my bag, and then came to this hotel overlooking the sea.

This was the first thing we saw as we came into the harbour. It is in a suburb called Nice Havrais, built by old Dufayel of Paris. It was a curious and pathetic sensation to see the Belgian flags still flying bravely. The different Ministries are set up here, and one villa has been set aside for the King and Queen, who have not yet left Belgian soil. The Legations are all established in this hotel and are bored to extinction, as their work has dropped very much. This little suburb enjoys all the privileges of extraterritoriality, and even the French Minister to Belgium goes through the motions of being accredited to a foreign Government in his country. The cars of the various Legations go buzzing around among the French and Belgian and British cars. The streets are full of troops of the three nations, while some twenty transports ride at anchor in the open roadstead. Fresh troops from England are arriving constantly, and march singing through the town to the camps outside, whence they are sent to the front. There are two British hospitals near this hotel—one of them the Casino—and wounded are everywhere. The place is astonishingly calm, but everybody knows there is a war. The French have their teeth set and are confident of the final outcome. Women are in the custom house, drive the trams, collect the fares and do a hundred other things that are usually out of their line.

I found the hall filled with colleagues, and exchanged greetings with the crowd before going over to the Foreign Office to make my bow. I found Colonel Fairholme packing, and ready to leave this evening for England.

The Foreign Office has a pretty little villa in a pretty little garden and keeps busy. I saw everybody, from Monsieur Davignon down to the porters, and spent an hour and a half there. Then at their request I went to the "Palace" and talked with General Jungbluth. He will try to arrange my business for me by telegraph, and will let me know in the morning whether I am to go up to the front to see the King and Queen.

When I came away from this call, Osborne was waiting for me and took me down to the Consulate for an hour's talk. Then back to the hotel to dine with Sir Francis. After dinner we all went out and bade the Colonel farewell.

Tuesday.—General Jungbluth was waiting for me when I came down this morning, to say that I should go to the front. Osborne was waiting with his car, and took me to the Ministry of War, to ask for a lift to Dunkerque in a military car. As luck would have it, to-day's car had left ten minutes before, so I was put off until to-morrow morning, when I shall go up with the W———s. I have spent a good part of the day getting my papers in order—both French and Belgian —and in the tiresome occupation of being photographed.

October 28th, Hôtel des Arcades, Dunkerque.—Another one-night stand.

We cleared out of Havre this morning over muddy and slippery roads. It rained hard all night, and we made good time by way of Fécamp, Dieppe, Eu, Abbeville, Montreuil, Bologne, Marquise, and Calais, getting to Dunkerque a little after four, just in time to smell the smoke of a couple of bombs dropped by an aeroplane across the street from the office of the Prime Minister, upon whom I called.

We began running into big bunches of troops at Abbeville—English, French and Belgian. I saw some of the Indian troops doing sentry duty and looking cold and uncomfortable, and did not blame them, for it was raw and cheerless. The Rolls-Royce is a beauty and sailed along all day like a gondola.

The Prime Minister had set up his office in the Mayor's room at the Hôtel de Ville, which I found in an uproar because of the bombs. The Prime Minister was said to be at Headquarters, at Furnes, across the Belgian frontier, and I was urged to go there to see him. We made twenty-one kilometers there, in time to find that little town in a great state of excitement, because three big shells had come from nobody knew where, and burst by the railroad station.

But the Prime Minister was not there, and it was dark, so we gathered up a guide and set off for la Panne, where the King and Queen are living. Neither of them was there; nobody but a gendarme on duty. The King was off with the troops and the Queen was looking after the wounded, who have overflowed all the hospitals. In the past week—just this one engagement—the Belgians have suffered 12,000 casualties.

The road from Furnes to la Panne and back lay close behind the lines, so that we could hear the steady roar of the fighting and see the bursting shells, particularly those from the British ships, which made a tremendous flash and roar.

We came on back to town, being stopped every minute by French outposts, and got to this hostelry at seven-thirty. While I was cleaning up, the Prime Minister came in and claimed me for dinner. He had his secretary, Count Lichtervelde, A. B., who is here looking after the wounded, and a couple of officers. And *then* we talked until the hands dropped off the clock and I was nearly dead for sleep. Then I took A. B. home to her hospital, through the streets darkened for the benefit of Count Zeppelin, and *now* I *am* ready for my rest.

I have plans for to-morrow, but shall see what happens to them when I see the Prime Minister in the morning.

October 29th.—Still at Dunkerque.

Another busy and interesting day, and if all goes well, I shall be back in London to-morrow night.

I was up early, did a little writing, and went over to see the Prime Minister, who was waiting for me. Despatched my business with him in short order, to my complete satisfaction. He is a trump, and it is a joy to do business with him, even at a time when he is hounded, as he is now.

He said the King was out with the troops, but had sent in to say he wanted to see me and would come in to headquarters at Furnes at four-thirty for that purpose. The Queen had also sent word in that she wanted to see me. She was busy looking after the wounded, but said she would come to la Panne at four. That suited me, although I was in some doubt as to how I would be able to make connections between the two audiences.

Last night I had talked of going out to look at the fighting, and A. B. had offered to conduct me. I had not taken the offer very seriously, but when I got back to the hotel after seeing the Prime Minister, she was there in a big racing car, with a crack chauffeur, ready for the jaunt. She was in her campaign kit of knickers, with a long rain-coat and a big knitted cap, and an entrancing boy she made. Mr. and Mrs. W——had asked to go along, and were in their car with Barbaçon, an aide-de-camp of the Prime Minister. Monsieur de Broqueville came out quite seriously and begged A. B. not to lead me into danger, whereat everybody had a good laugh.

We made quick time to Furnes and drew up before Headquarters, where we learned what was known of the lay of the land and the points of the front we could reach without getting in the way. The Belgians, who had for ten days held the line of the Yser from Nieuport to Dixmude, waiting for reinforcements to come up, had been obliged to fall back to the line of the railroad, which forms the chord of the arc, and had inundated the intervening territory to impede the German advance. French and English troops were being brought up in large numbers to relieve the Belgians, who have lost in killed and wounded nearly a third of the 50,000 men engaged.

While waiting for some definite news to be brought in for us, we climbed to the top of the high tower of the market next the Hôtel de Ville, for a look at the battle line. It was pretty misty, but we could see the smoke of shrapnel and of the big shells from the English ships, which were enfilading the German right.

The staircase up this tower was a crazy thing, with rotten steps and places where two or three steps were missing altogether. It was bad enough going up where we could take hold and pull ourselves up, but it was far worse going down, because we were ordered down in a hurry and all came piling down in a steady stream. There were squeaks and screams at the bad moments, but we did manage to get down without mishap and take stock of ourselves.

We found some German prisoners lying on the straw in the entrance hall, and stopped to speak to them. They said that their troops were very tired from long, hard fighting, but that they had plenty of men. They seemed rather depressed themselves.

By the time we got down, our information had come and we set off through a welter of transport trains, artillery, ambulances, marching troops, and goodness knows what else, in the direction of X——. When we got within a couple of kilometers of the place, an officer stopped us and asked if we knew where we were going. He shrugged his shoulders when we said we did, and let us go straight into it. When we were bowling along about one kilometer from the town, three shells burst at once, about two hundred yards to our left, and we stopped to see what was toward. A hundred yards ahead to the right of the road was a battery of five big guns, and the Germans were evidently trying to get their range. The shells kept falling to the left, near a group of farm-houses, and as some of the spent balls of shrapnel kept rolling around near us, we decided we might as well go and see the big guns from nearer to.

In the shelter of the farm-houses were fifty or sixty men, some of them cooking their lunch, others sleeping, all quite oblivious of the roar of bursting shrapnel and the spattering of the bullets near by. And a few months ago probably any of these men would have been frightened into a fit by a shell bursting in his neighbourhood. It is wonderful how soon people become contemptuous of danger. The horses that were tethered by the roadside seemed to take it all as a matter of course, and munched away at their hay, as though all the world were at peace. A wobbly cart came creaking by with an infantryman, who had had a good part of his face shot away. He had been bandaged after a fashion and sat up blinking at us stupidly as the cart lumbered by, bumping into holes and sliding into ruts.

I was not keen on staying longer than was necessary to see what was there, but W.——was very deliberate and not to be budged for more than half an hour. We finally got him started by calling his attention to the spent balls, which make a tremendous singing noise, but do no harm. The only really safe thing in the neighbourhood was what did the trick. The Germans were making a furious attack, evidently determined to break the line before the fresh troops could be brought up, and the cannonading was terrific. The whole front as far as we could see in either direction was a line of puffs of smoke from bursting shrapnel and black spouts of earth from exploding shells. The crackle of the *mitrailleuses* rippled up and down the whole line. The Belgians were pounding back as hard as they could and the noise was deafening. Finally, when we decided to leave, the officer in command of the battery loaded all five guns at once and fired a salvo for our benefit. The great shells tore away, roaring like so many express trains, and screaming like beasts in agony—a terrifying combination. My ears ache yet. It was getting hotter every minute and the Germans were evidently getting a better idea of the range, for the shells began falling pretty close on the other side, and I was quieter in my mind when we went back to our cars and pulled out of the actual line. We took a road a few hundred yards back, parallel with the lines, and drove along slowly, watching the effect of the shell fire, until we absolutely had to start back for lunch. On the way we stopped at a peasant's hut, and said hello to Jack Reyntiens.

When we got back to the hotel, about half an hour late for lunch, we found the Prime Minister waiting for us. At the door, in addition to the usual sentry, there were two privates of the *chasseurs à cheval*, one wearing a commander's star of the Legion of Honor. They saluted and smiled, and I bowed and went on in to my meal. They came in after me, still smiling, and I was taxed with not recognising them. They were the Duc d'Ursel and ———, the heads of their respective houses, who had enlisted, and are still fighting as privates. They had just been relieved and were on their way to the rear, where the Belgian army is being reformed and rested.

As soon as we had got through, I had to start back for my audience of the Queen. W.——took me out to la Panne, where we found the Villa on the sand dunes, a little way back of the lines. There were a couple of gendarmes on duty, the King's Secretary, and the Countess de Caraman-Chimay, the one Lady-in-Waiting. I had just got inside when the door opened and the King came in. He had heard I was coming to see the Queen and had motored down from Furnes. I was able to satisfy him in a few minutes on the points he had wanted to see me about and then he questioned me about friends in Brussels. I suggested to him that it would probably help our committee in raising funds if he would write an appeal for help from America. He fell in with the idea at once, and together we got out an appeal that is to be sent across the water. Where we sat we could see the British ships shelling the Germans, and the windows of the dining-room were rattling steadily. The King stood beside the table with his finger tips resting on the cloth, watching the stuff ground out word by word. I looked up at him once, but could not bear to do it again—it was the saddest face one can imagine, but not a word of complaint was breathed.

Just as we were finishing, the Queen came and bade us in to tea. She was supposed to wait for her Lady-in-Waiting to bring me, but didn't. The King stayed only a minute or two and then said he must be getting back to Headquarters, where he would see me later.

I suggested to the Queen that she, too, make an appeal to the women of America, to which she agreed. Another appeal was prepared for her, and it, too, will be sent to America by the first post.

The Queen had wanted to see me about the subject of surgeons for the Belgian army. The Belgian surgeons in the Brussels hospitals have been replaced by

Germans, and have nothing to do, although they are desperately needed here. The Queen was terribly depressed about the condition of the wounded. There are so few surgeons, and such tremendous numbers of wounded, that they cannot by any possibility be properly cared for. Legs and arms are being ruthlessly amputated in hundreds of cases where they could be saved by a careful operation. Careful operations are, of course, out of the question, with the wounded being dumped in every minute by the score. In these little frontier towns there are no hospital facilities to speak of, and the poor devils are lucky if they get a bed of straw under any sort of roof, and medical attendance, within twenty-four hours. We went to see one hospital in a near-by Villa, and I hope I shall never again have to go through such an ordeal. Such suffering and such lack of comforts I have never seen, but I take off my hat to the nerve of the wounded, and the nurses, most of them the best class of Belgian women, used to every luxury and getting none.

The Queen gave me tea, and one of her small supply of cigarettes, and we talked until after dark. The monitors off shore had been joined by a battleship, and the row was terrific and rendered conversation difficult.

The Queen was still full of courage and said that as long as there was one square foot of Belgian soil free of Germans, she would be on it. She said it simply, in answer to a question from me, but there was a big force of courage and determination behind it. As I was not dismissed, I finally took it on myself to go, and the Queen came with me to the door and sent me on my way. She stood in the lighted doorway until

I reached the motor, and then turned slowly and went in—a delicate little woman with a lion's heart. Inglebleek and the Countess de Caraman-Chimay came out after we had cranked the car, and gave me messages for their families and friends. It is a pretty hard change for these people, who three months ago were leading such a dull, comfortable life, but they have risen to it with fine spirit.

The King was with his staff, studying the maps and despatches, when I got to Furnes, and I was shown the whole situation—most interesting on the large scale maps that show every farm-house and pathway. I was to go back to Dunkerque with Monsieur de Broqueville, so waited while they discussed the events of the day and plans for to-morrow.

While they talked reinforcements were pouring through the town, with great rumbling of artillery and blowing of trumpets. It was a comforting sound, as it presaged some relief for the Belgians in their heart-breaking stand.

There was comfort in riding back through the night with the Prime Minister, for there was no long examination of papers, etc. When we came to a post, the aide-de-camp would switch on a strong light in the car, the sentries would salute, and on we would go at a great gait.

Seemingly I was boarding with Monsieur de Broqueville, as I was led back to dine with him.

To-morrow I am off to London. Loewenstein, a young Brussels banker, is to take me over in his racing car, which is a useful institution these days. We take along his mother-in-law, Madame Misonne, and A. B.

It means getting up at five to motor to Calais to catch the boat. There the car will be slung aboard, so that we can be whisked up to London without waiting for a train.

On board S.S. "Orange Nassau," North Sea, November 2, 1914.—On Friday morning we were called before dawn, and got under way as per schedule— Loewenstein, Madame Misonne, A. B., and I. We made good time, over slippery roads, to Calais, despite frequent stops to have our papers examined by posts, and got to the dock some twenty minutes before the steamer sailed. The car was hoisted aboard, and we rode across in it. Frederick Palmer was on board, returning in disgust after having been just that far toward the front.

Our suicide wagon was swung off onto the dock without loss of time, and we sped away toward London while our fellow-passengers were doomed to wait for all sorts of formalities. It was a wild ride. At times we were doing as high as one hundred and thirty kilometers an hour over winding English roads, and I was somewhat relieved when I was dropped at the Embassy, safe and sound.

I got off some telegrams about my trip, and was told the Ambassador wanted to see me. Hoover was with him, and I turned over to them the appeals from the King and Queen.

Jack Scranton decided to come back to Brussels with me, to give me a hand in Legation work, and spent the morning packing enough plunder to see him through a siege of three or four years. A. B. came on to London to see her brother who is seriously wounded and in hospital. Now her family want her to return to Brussels and have placed her in my care for the journey.

This morning we had a crowd at the station to see us off. Countess N.——has also come along, and was entrusted to our care. A. B.'s family was there in force to say good-bye, so altogether the casual observer might have inferred that we were popular.

Brussels, November 5th.—We were met in Flushing by our Consular Agent, who put us through the customs and onto the train.

No motor was waiting for us at Rosendaal, and we had a hard time getting shelter for the night. Finally we succeeded in getting a room for the two women in a little, third-rate hotel, and Jack and I slept on the floor of a sitting-room in the little Hôtel Central. I was so dog-tired that I slept like a log, wrapped up in my fur coat.

While we were having coffee, M. de Leval came up in my little car. He had been to Rotterdam in connection with the first shipment of food, and thought he would find me alone. He had bought a lot of gasoline in Breda, to be called for, so we could take no luggage. We found another car leaving for Brussels at noon, and loaded it up with Countess N.,——Jack and the luggage, while M. de L. and I took A. B. and the mail bags, and started by way of Breda. We came through Aerschot and stopped for a stretch and to look about.

We walked about the streets for a time, and stopped in a shop to ask for a drink of water. After giving it to us, the proprietor asked if we would like to see the

An improvised pass, written on the back of a wine card, to
enable Mr. Gibson to pass through the streets after dark on
his way to German headquarters at Liège

Pho'ograph by Paul Thompson

General Baron von Bissing, Governor-General in Belgium

state the Germans had left things in. He led us back into his living quarters, opened a door bearing an inscription to the effect that it was an officers' mess, and let us in. I never have seen a more complete mess. Everything in the place was smashed, and the whole room was filthy. The officers had left only a few days before and had taken pains to break everything before they went. Obscene remarks were chalked on the walls, and the pictures were improved with heavy attempts at fun. I always used to think that the term "officer and gentleman" was redundant, but now I begin to understand the need for it.

The church was also in a bad state. The doors have nearly all been battered down. The wooden Gothic statues in the nave have been smashed or destroyed by fire. The altars and confessionals were wantonly destroyed. The collection boxes had been pried open and emptied. We were told that the holy-water font and the vestments of the priests had been profaned and befouled. It is not a pretty sight.

Aerschot was partially destroyed on August 19th and 20th. The Germans claim that their commanding officer was shot by the son of the Burgomaster. The Belgians claim that he was struck by a stray bullet fired at random by one of his own men in the marketplace. However that may be, the whole place was instantly in an uproar, and quiet was not restored until the town had been sacked and over one hundred and fifty people killed, among them women and children. The Burgomaster and his son and a priest were among those shot and buried outside the Louvain gate. One of those taken to the place of execution was spared on condition that he should go to Louvain to tell of what had happened.

Louvain has been cleaned up a lot, and we stopped there only long enough to have our passes examined at Headquarters, getting back a little before six to a warm welcome.

The other motor was due at six, but did not come, and after waiting up till midnight, I turned in. Jack bobbed up yesterday at noon. The car had been stopped at the frontier because several of the passengers had not proper papers. Jack threw out his chest and insisted on being taken to Antwerp to see the Military Governor. His passport, as bearer of despatches, did the business, and they were allowed to proceed under armed guard. They were kept overnight in the Hôtel Webber, and then Jack and Mme. N—— were allowed to come on to Brussels in the car, while the others were detained.

Marshal Langhorne came in to-day from The Hague to effect formal delivery of the first bargeload of food, and had weird tales to tell of his adventures by the way. Thank goodness, the first of the food has arrived in time, and if the flow can be kept up, the worst of our troubles will be averted.

With this first consignment of food came the story of how it was got through in such record time. Hoover is one of these people who is inclined to get things done and attend later to such details as getting formal permission, etc.

With Shaler's forty thousand pounds and promises of five hundred thousand dollars more, he went to work and placed orders for twenty thousand tons of food, costing two million dollars a week. This he did on the theory that money would come along later, when the need was realised, but that the Belgian stomachs would not wait until collections had been made. He purchased the food, got it transported to the docks, and loaded on vessels that he had contrived to charter, while all the world was fighting for tonnage, got them loaded and the hatches closed.

When everything was ready, Hoover went to the proper authority and asked for permission to ship the food, announcing that unless he could get four ship-loads of food into Belgium by the end of the week, the people would begin to starve. The functionary was sympathetic, but regretted that in the circumstances, he could not help. It was out of the question to purchase food. The railways were choked with troops, munitions and supplies. Ships were not to be had for love or money. And above all, the Channel was closed to commerce.

Hoover heard him patiently to the end.

"I have attended to all this," he said. "The ships are already loaded and ready to sail. All I need from you is clearance papers. You can let me have them, and everything will be all right."

The high official could hardly believe his ears:

"Young man," he gasped, "perhaps you don't realise what you have done. Men have been sent to the Tower for less. If it were for any other cause, I hesitate to think what would happen to you. But as it is, I can only congratulate you on some very good work."

And that's how we got our food in time.

Fines are being imposed on towns on one pretext or another. The other day two policemen got into a controversy with a German secret-service agent who did not explain who he was, and got a good thumping for doing various things that a civilian had no business to do. This morning von Lüttwitz comes out with this proclamation:

On the 28th of October, 1914, a legally constituted court martial pronounced the following sentences:

(1) The policeman De Ryckere for having attacked, in the legal exercise of his duties, an authorised agent of the German Government, for having deliberately inflicted bodily hurt in two instances with the aid of other persons, for having aided in the escape of a prisoner and for having attacked a German soldier, was condemned to five years' imprisonment.

(2) The policeman Seghers for having attacked, in the exercise of his legal duties, an authorised agent of the German Government, for having deliberately inflicted bodily hurt on this German agent, and for having aided the escape of a prisoner (all these offences constituting one charge), was condemned to three years' imprisonment.

The sentences were confirmed on October 31st by the Governor-General, Baron von der Goltz.

The city of Brussels, not including its suburbs, has been punished for the injury by its policeman De Ryckere to a German soldier, by an additional fine of Five Million Francs.

The Governor of Brussels,

Brussels, November 1, 1914.   BARON VON LÜTTWITZ,
General.

Last night we dined at Ctesse. N——'s to celebrate everybody's safe return.

Brussels, Sunday, November 8, 1914.—Barges of food are beginning to come in, and we have the place filled with people with real business concerning the food and a lot of the usual "halo-grabbers" anxious to give advice or edge into some sort of non-working position where they can reap a little credit.

We are put on German time to-day.

On November 4th the Governor-General came out with a proclamation ordering that German money be accepted in all business transactions. It is to have forced currency at the rate of one mark to one franc, twenty-five centimes. As a matter of fact, it is really worth about one franc, seven centimes, and can be bought at that rate in Holland or Switzerland, where people are glad enough to get rid of their German money. Any shop refusing to accept German paper money at the stipulated rate is to be immediately closed, according to the Governor's threat.

Brussels, November 9, 1914.—Late in the afternoon Jack and I took Max for a run in the Bois. While we were going across one of the broad stretches of lawn, an officer on horseback passed us, accompanied by a mounted orderly. To our surprise the orderly drew his revolver and began waving it at us, shouting at the same time that if that —— —— —— dog came any nearer, he would shoot him down. The officer paid no attention, but rode on ahead. I started after them on foot, but they began to trot and left me in the lurch. I ran back to the motor, overtook them, and placed the car across their path. The officer motioned his orderly to go ahead, and then let me tackle him. He took the high ground that I had no reason to complain since the dog had not actually been shot, not seeming to realize that peaceable civilians might have legitimate objections to the promiscuous waving of revolvers.   He declined to give his name or that of the soldier, and I gave up and let him ride on after expressing some unflattering opinions of him and his kind to the delight of the crowd that had gathered.   They did not dare say anything direct, but as I got back into the car they set up a loud *"Vive l'Amérique."*   The officer looked peevish and rode away very stiff and haughty.   Of course, since he refused to give his name, there was no getting at him, and I was free to be as indignant as I liked.

The Germans are tightening up on the question of travel in the occupied territory, and we are now engaged in a disagreeable row with them over passes for the Legation cars.   They want to limit us in all sorts of ways that make no difference to them, but cut down our comfort.   They will probably end by giving us what they want; but when it is all done we shall have no feeling of obligation, having been forced to fight for it.

Brussels, November 14, 1914.—On the morning of the 10th, I came down to the Legation and found things in an uproar.   A telegram had been received saying that two trainloads of food, the first shipment for the Province of Liège, would cross the frontier in the course of the afternoon, under convoy of Captain Sunderland, our Military Attaché at The Hague.   The Minister and I are the only people authorized to receive shipments; and, as no power of attorney had been sent to the Consul at Liège, things were in a nice mess; and, at the request of the German authorities and the

Committee, it was decided that I should go down, receive the stuff and make arrangements for its protection and for the reception of future shipments. The German authorities were so excited about my being there to head off any trouble that they hustled me off on an hour's notice without any lunch. I contrived to get Jack's name put on the *laisser-passer*, so that he could go along and see a little something of the country. Joseph, the Legation butler, was wild to go along as far as his native village to see his aged ma, whom he had not seen since the beginning of the war, and he rode on the front seat with Max who was much delighted to get under way again.

Jack was thrilled with the trip, and nearly fell out of the car going through Louvain and the other ruined villages along the way. As we were in such a rush, I could not stop to show him very much; but in most of these places no guide is needed. Louvain has been cleared up to a remarkable extent, and the streets between the ruined houses are neat and clean. On my other trips I had had to go around by way of Namur, but this time we went direct; and I got my first glimpse of Tirlemont and St. Trond, etc.

When we reached Liège we went straight to the Consulate without pausing to set ourselves up at a hotel, but found that nothing was known of Captain Sunderland or his food trains. Thence to the German headquarters where we inquired at all the offices in turn and found that the gentleman had not been heard from. By the time we got through our inquiries it was dark; and, as we had no *laisser-passer* to be out after dark, we had to scuttle back to the hotel and stay.

In the morning the Consul and I started off again to
see what had become of our man. We went through
all the offices again, and as we were about to give up,
I found Renner, who used to be Military Attaché of
the German Legation here, and is now Chief of Staff
to the Military Governor. He cleared up the mystery.
Sunderland had arrived about the same time I did, but
had been taken in hand by some staff officers, dined
at their mess, and kept busy until time for him to be
off for Maestricht. He was, however, expected back
in time to lunch at the officers' mess. He was also
expected to dine with them in the evening. I left word
that I wanted to see him and made off to get in touch
with the members of the local committee and make
arrangements as to what was to be done with the food.
We sat and waited until nearly dark, when I decided
to go out for a little spin. I gathered Jack and the
Consular family into the car and went for a short
spin.

After losing our way a couple of times we brought up
at the Fort of Chaudefontaine, which was demolished
by the Germans. It is on top of a veritable mountain
and it took us some time to work our way up on the
winding road. When we got there the soldiers on
guard made no trouble and told us that we could mouse
around for fifteen minutes. We walked out to the
earthworks, which had been made by the Belgians and
strengthened by the Germans, and then took a look
at the fort itself, which was destroyed, and has since
been reconstructed by the Germans. They must have
had the turrets and cupolas already built and ready
to ship to Liège, for the forts are stronger than they

ever were before and will probably offer a solid resistance when the tide swings back, unless, of course, the allies have by that time some of the big guns that will drop shells vertically and destroy these works the way the German 42's destroyed their predecessors. It was very interesting to see and hard to realise that up to three months ago this sort of thing was considered practically impregnable.

When we got back we found that our man had come and had left word that he could be found at the Café du Phare at six o'clock. We made straight for that place, and found him. I made an appointment with him for the first thing next morning, and went my way.

I was bid to dine with the German Military Governor and his staff, but told Renner that since we were accredited here to the Belgian Government, accepting German hospitality would certainly be considered as an affront. He saw the point, and did not take offence, but asked me to come over after dinner for a talk and bring Jack along, the which I promised to do. While we were dining, a soldier with a rifle on his shoulder strode into the dining-room and handed me a paper; great excitement, as everybody thought we had been arrested. The paper was a pass for us to circulate on the streets after dark, so that we could go over to the headquarters. It was written on the back of a menu in pencil. Although dinner was over the entire mess was still gathered about the table discussing beer and Weltpolitik. At the head of the table was Excellenz Lieutenant-General von Somethingorother, who was commanding a German army on the eastern front when they got within fifteen miles of Warsaw. After being driven back he had an official "nervous break-down," and was sent here as Governor of the Province of Liège—quite a descent, and enough to cause a nervous breakdown. There was another old chap who had fought in the Franco-Prussian war and had not yet quite caught up with this one. I foregathered with Renner and got my shop talk done in a very short time. Then everybody set to to explain to us about the war and what they fought each other for. It was very interesting to get the point of view, and we stayed on until nearly midnight, tramping home through a tremendous downpour, which soaked us.

The next morning at eleven I met Sunderland. We saw the Governor and the Mayor and Echevins, and talked things out at length. I had to collect a part of the cost of the food before I could turn it over, and they explained that the chairman of the local committee had gone to Brussels to negotiate a loan; he would be back in four or five days and if I would just wait, they would settle everything beautifully. That did not please me, so I suggested in my usual simple and direct way that the Governor rob the safe and pay me with provincial funds, trusting to be paid later by the committee. It took some little argument to convince him, but he had good nerve, and by half-past twelve he brought forth 275,000 francs in bank-notes and handed them over to me for a receipt. Sticking this into my pocket, I made ready to get under way, but there was nothing for it but that I must lunch with them all. Finally I accepted, on the understanding that it would be short and that I could get away immediately afterward. That was not definite enough, however, for we sat at table until four o'clock and then listened to some speeches.

When we got down the home stretch, the Governor arose and made a very neat little speech, thanking us for what we had done to get food to the people of Liège, and expressing gratitude to the American Government and people, etc.   I responded in remarks of almost record shortness, and as soon as possible afterward, we got away through the rain to Brussels.

After getting through that elaborate luncheon, getting our things ready at the hotel, paying our bill, saying good-bye all around once more, etc., it was nearly five o'clock when we got off and nearly eight when we reached Brussels and put our treasure in the safe.

The Germans have begun arresting British civilians and we have had our hands full dealing with poor people who don't want to be arrested and kept in prison until the end of the war and can't quite understand why *they* have to put up with it.   It is pretty tough, but just another of the hardships of the war, and while we are doing our best to have the treatment of these people made as lenient as possible, we can't save them.

Brussels, November 16, 1914.—Some more excitement yesterday morning, when various British subjects were arrested.

Two German civilians tried to force their way into the British Consulate and arrest Mr. Jeffes, the British Consul, and his son, although the American flag was flying over the door and there was a sign posted to the effect that the place was under our protection and all business should be transacted with us. Fortunately Nasmith was there, and after trying to explain the matter politely, he made for the two men, threw them into the street, and bolted the door. The gum-shoe men were so surprised that they went away and have not been back. Last night I was called around to the Consulate and found two more men shadowing the place. There seemed to be no danger of arrest, but Nasmith spent the night there, and this morning I went around and took the Jeffes to our Consulate, so that if any attempt was made to take them, we should have an opportunity to protest. The higher authorities had promised not to seize them, but apparently you can never tell.

Yesterday was the King's Saint's Day, and word was passed around that there would be a special mass at Ste. Gudule. Just before it was to begin, the military authorities sent around and forbade the service. The Grand Marshal of the Court opened the King's book at his house, so that we could all go around and sign, as in ordinary times, for we are accredited to the King of the Belgians, but early in the morning an officer arrived and confiscated the book. The Government of Occupation seems to be mighty busy doing pin-head things for people who have a war on their hands.

Countess de Buisseret's little boy was playing on the street yesterday when the German troops passed by. Being a frightful and dangerous criminal, he imitated their goose-step and was arrested. M. de Leval went around to headquarters to see what could be done, supposing, of course, that when it was seen what a child he was, his release would be ordered.   Instead, he was told seriously that the youngster must be punished and would be left in jail for some days.

Brussels, November 18, 1914.—This is another day of disgust.   This morning one of the servants of the Golf Club came in to say that there were fifty German soldiers looting the place.   In the afternoon Jack and I went out for a look at the place and to get my clubs. We found a lot of soldiers under command of a corporal.   They had cleaned the place out of food, wine, linen, silver, and goodness knows what else.   Florimont, the steward, had been arrested because he would not tell them which of the English members of the club had gone away and where the others were staying.   Having spent his time at the club, the fact was that he did not know who was still in town and could not tell, but the Germans could not be convinced of this and have made him prisoner.

I stopped at headquarters this afternoon to see von der Lancken.   As I came out a fine Rolls-Royce limousine drew up on the opposite side of the street—a military car.   The chauffeur, in backing out, caught and tore the sleeve of his coat.   In a rage, he slammed the door and planted a tremendous kick in the middle of the panel with his heavy boot.   I stood agape and watched.   He looked up, caught me looking at him, and turned his anger from the motor to me.   He put his hands on his hips, shot out his jaw and glared at me.   Then he began walking toward me across the street in heavy-villain steps, glaring all the time.   He stopped just in front of me, his face twitching with rage, evidently ready to do something cataclysmic. Then the heavens opened, and a tremendous roar came from across the street. The officer to whom the car belonged had seen the display of temper from his window, and had run out to express his views. The soldier did a Genée toe-spin and stood at attention, while his superior cursed him in the most stupendous way. I was glad to be saved and to have such a display of fireworks into the bargain.

November 19th.—One day is like another in its cussedness.

The Germans have been hounding the British Legation and Consulate, and we have had to get excited about it. Then they announced to the Dutch Chargé that our courier could no longer go—that everything would have to be sent by German field post. You would think that after the amount of hard work we have done for the protection of German interests and the scrupulous way in which we have used any privileges we have been accorded, they would exert themselves to make our task as easy as possible and show us some confidence. On the contrary, they treat us as we would be ashamed to treat our enemies.

This morning it was snowing beautifully when I woke up, a light, dry snow that lay on the ground. It has been coming down gently all day and the town is a lovely sight, but I can't get out of my mind the thought of those poor beggars out in the trenches. It seems wicked to be comfortable before a good fire with those millions of men suffering as they are out at the front.

And now Grant-Watson* has been put in prison. He stayed on here after the Minister left, to attend to various matters, and was here when the Germans arrived. Recently we have been trying to arrange for passports, so that he and Felix Jeffes, the Vice-Consul, might return to England. The authorities were seemingly unable to make up their minds as to what should be done, but assured the Minister that both men would be allowed to return to England or to remain quietly in Brussels. On Friday, however, the Germans changed their minds and did not let a little thing like their word of honour stand in the way.

The Minister was asked to bring Grant-Watson to headquarters to talk things over—nothing more. When they got there, it was smilingly announced that Grant-Watson was to leave for Berlin on the seven o'clock train, which put us in the position of having lured him to prison. The Minister protested vigorously, and finally Grant-Watson was put on parole and allowed to return to the Legation, to remain there until eleven o'clock yesterday morning. I went over the first thing in the morning to help him get ready for his stay in jail. At eleven Conrad arrived in a motor with Monsieur de Leval. We went out and got in, and drove in state to the École Militaire, and, although I was boiling with rage at the entire performance, I could not help seeing some fun in it.

Grant-Watson's butler was ordered to be ready to go at the same time. At the last minute the butler came down and said perfectly seriously that he would

*Second Secretary of the British Legation in Brussels.

not be able to go until afternoon, as he had broken the key to his portmanteau and would have to have another made. The Germans did not see anything funny in that, and left him behind.

When we got to the École Militaire, we were refused admittance, and had to wrangle with the sentries at the door. After arguing with several officers and pleading that we had a man with us who wanted to be put in prison, we were reluctantly admitted to the outer gate of the building, where British subjects are kept. When the keeper of the dungeon came out, I explained to him that the butler had been detained, but would be along in the course of the afternoon, whereupon the solemn jailer earnestly replied, "Please tell him that he must be here not later than three o'clock, or he can't get in!" And nobody cracked a smile until I let my feelings get the better of me.

I was prepared for an affecting parting with Grant-Watson in consigning him to the depths of a German jail, but he took it as calmly as though he were going into a country house for a week-end party. I suppose there is some chance that they may exchange him for a few wounded German officers and thus get him back to England.

Since our snow-storm the other day, the weather has turned terribly cold and we have suffered even with all the comforts that we have. And the cheerful weather prophets are telling us that without doubt this will be one of the coldest winters ever known. A pleasant prospect for the boys at the front! Mrs. Whitlock and everybody else is busy getting warm

Photographs of Dinant

View of Dinant

clothing for the poor and for the refugees from all parts of Belgium who were unable to save anything from their ruined homes. It is bad enough now, but what is coming. . . .

Gustave has just come in with the cheering news that Ashley, our crack stenographer, has been arrested by the Germans. They are making themselves altogether charming and agreeable to us.

Max is spread out before the fire, snoring like a sawmill—the only Englishman in Brussels who is easy in his mind and need not worry.

Tuesday, November 24th.—Another day of rush without getting very far.

The Germans decided this morning that they would arrest Felix Jeffes, the British Vice-Consul, so I had the pleasant task of telling him that he was wanted. I am to go for him to-morrow morning and take him to the École Militaire with his compatriots. This job of policeman does not appeal to me, even if it is solely to save our friends the humiliation of being taken through the streets by the Germans.

November 25th.—Had a *pleasant* day.

Had arrangements made with Jeffes to go with him to the École Militaire at 11 o'clock and turn him over to his jailer. The Minister went up with von der Lancken to see the Englishmen and be there when Jeffes arrived, so as to show a friendly interest in his being well treated.

I went around to the Consulate on time, and found

that, through a misunderstanding, Jeffes had made no preparations for going, having been assured that another attempt would be made to get him off. I pointed out that the Minister had given his word of honour that Jeffes should be there, and that he would be left in a very unpleasant and annoying position if we did not turn up as promised. Jeffes was perfectly ready, although not willing to go. I went to the Ecole Militaire and explained to von der Lancken that Jeffes' failure to appear was due to a mistake, and asked that he be given time to straighten out his accounts and come later in the day or to-morrow morning. The answer was that he must come some time during the day. The Consul-General went straight to von Lüttwitz with Jeffes, made a great plea on the score of his health or lack of it, and got his time extended until he could be given a medical examination by the military authorities. Late in the afternoon he was looked over and told to go home and be quiet, that he would probably not be wanted, but that if anything came up, they would communicate with him further.

Brussels, November 27, 1914.—More busy days. Each day we swear that we will stop work early and go out to play. Each day we sit at our desks, and darkness comes down upon us, and we do not get away until nearly eight o'clock. "Thanksgiving Day" was no exception, and to-day we are going through the same old performance. Yesterday, by strenuous work, I got down to swept bunkers and had a good prospect of an easy day. Instead of that there has been a deluge of

Consuls, mail, telegrams, and excited callers, and we are snowed under a heap of work it will take several days to get out of the way.

We came back to them with a bump, however, when Nasmith came to my flat at midnight to say that Jeffes had been arrested. And it was done in the usual charming manner. In the course of the afternoon, the Consul-General got a note asking him to go to headquarters "to talk over the case of Mr. Jeffes." It asked also that Jeffes accompany the Consul-General "to the conference." When they arrived it was announced that Jeffes was under arrest and to be sent immediately to the Ecole Militaire. The Consul-General, like the Minister, on the occasion of his visit, was placed in the position of having lured his friend into jail. He protested vigorously, but was not even allowed to accompany Jeffes to the Ecole Militaire. It was only after some heated argument that Jeffes was allowed five minutes at home, under guard, to get a few belongings together to take with him. The Consul-General is furious, and so am I when I remember how decently the German Vice-Consul here was treated when the war broke out.

Early in the week Jack is to be sent down to Mons, to bring out some English nurses who have been there nursing the British wounded. Two of them, Miss Hozier and Miss Angela Manners, were in yesterday. They have been working hard during the past three months and are now ready to go back to England if we can arrange for passports.

Under the date of November 26th, General von Kraewel announces that he has succeeded Baron von

Lüttwitz, who has been transferred to the army at the front.

Hoover arrived from London this afternoon accompanied by Shaler and by Dr. Rose, Henry James, Jr., and Mr. Bicknell of the Rockefeller Foundation, who have come to look into conditions. There is plenty for them to see, and we shall do our best to help them see it.

As we learned from a confidential source, several days ago, there has been a big shake-up in the Government here. Both von der Goltz and von Lüttwitz have gone and have been replaced—the first by Freiherr von Bissing, and the latter by General von Kraewel. There are several explanations for the changes, but we don't yet know what they mean.

Brussels, December 2, 1914.—We have had a hectic time. Hoover arrived on Sunday evening, accompanied by Shaler and by three representatives of the Rockefeller Foundation. We have had a steady rush of meetings, conferences, etc., and Hoover and Shaler pulled out early this morning. There is not much relief in sight, however, for tomorrow morning at the crack of dawn, I expect to start off on a tour of Belgium, to show the Rockefeller people what conditions really are. We shall be gone for several days and shall cover pretty well the whole country.

Yesterday morning I got Jack off to Mons to bring back the British nurses. Everything in the way of passports and arrangements with the military authorities had been made, and he went away in high spirits for a little jaunt by himself. This morning at half-past

three o'clock he rang the doorbell and came bristling
in, the maddest man I have seen in a long time. He
had suffered everything that could be thought of in
the way of insult and indignity, and to make it worse,
had been obliged to stand by and watch some brutes
insult the girls he was sent down to protect. When he
arrived at Mons he got the nurses together and took
them to the headquarters, where he explained that he
had been sent down by the Minister with the consent
of the German authorities, to bring the nurses to Brussels. This was stated in writing on the passport given
him by the German authorities here. Instead of the
polite reception he had expected, the German officer,
acting for the Commandant, turned on him and told
him that the nurses were to be arrested, and could not
go to Brussels. Then, by way of afterthought, he
decided to arrest Jack and had him placed under guard
on a long bench in the headquarters, where he was kept
for three hours. Luckily, an old gentleman of the
town who knew the nurses, came in on some errand,
and before they could be shut up, they contrived to tell
him what the situation was and ask him to get word
to the Legation. Right away after this the three
women were taken out and put in the fourth-class cells
of the military prison, that is, in the same rooms with
common criminals. Jack was left in the guard room.
The old gentleman, who had come in, rushed off to the
Burgomaster and got him stirred up about the case,
although he was loath to do anything, as he *knew* that
a representative of the American Legation could not
be arrested. Finally he did come around to headquarters, and after a long row with the Adjutant, they

got Jack released and fitted out with a *laisser-passer* to return to Brussels. He was insulted in good shape, and told that if he came back again, sent by the Minister or by anybody else, he would be chucked into jail and stay there. Before the nurses were taken down to their prison, the Adjutant shook his fist in Miss Hozier's face, and told her that they were going to give her a good lesson, so that the English should have a taste of the sort of treatment they were meting out to German nurses and doctors that fell into their hands.

The Mayor and Aldermen took Jack in charge when he was released, and kept him in one of their homes until time for the train to leave for Brussels at midnight. They were convinced that he would be arrested again at the station, but he did get off in a car filled with sick soldiers and arrived here without mishap at three o'clock or a little after.

I went over to see von der Lancken the first thing in the morning, and told him the whole story, in order that he might be thinking over what he was going to do about it before the Minister went over to see him at eleven. The Minister said his say in plain language, and got a promise that steps would be taken at once to get the girls out of prison and have them brought to Brussels. Later in the day von der Lancken came through with the information that the action of the authorities at Mons was "*due to a misunderstanding*," and that everything was lovely now. We suppose that the girls will be here to-morrow; if not, inquiries will be made and the Minister will probably go down himself.

Yesterday morning we spent visiting soup kitchens, milk stations, and the distributing centres for supplying old clothes to the poor. The whole thing is under one organisation and most wonderfully handled. It is probably the biggest thing of the sort that has ever been undertaken and is being done magnificently.

It is a curious thing to watch the Commission grow. It started as nothing but a group of American mining engineers, with the sympathetic aid of some of our diplomatic representatives and the good-will of the neutral world. It is rapidly growing into a powerful international entity, negotiating agreements with the Great Powers of Europe, enjoying rights that no Government enjoys, and as the warring governments come to understand its sincerity and honesty, gaining influence and authority day by day.

There is no explanation of the departure of von der Goltz. His successor has come out with a proclamation in three lines, as follows:

His Majesty, the Emperor and King, having deigned to appoint me Governor-General in Belgium, I have to-day assumed the direction of affairs.

BARON VON BISSING.

Brussels, December 3, 1914.

Brussels, Sunday, December 6, 1914.—We got away at eight o'clock on Thursday morning, in three cars from the Palace Hotel. We were four cars when we started, but fifty feet from the door the leading car broke down and could not be started, so we rearranged ourselves and left the wreck behind. The party was composed of the three Rockefeller representatives, Dr. Rose, Mr. Bicknell, and Henry James, Jr., Monsieur Francqui, Josse Allard, Jack and I.

It was rainy and cold, but we made good time to Louvain and stopped at the Hôtel de Ville. Professor Neerincxs, of the University, took up the duties of Burgomaster when the Germans shipped the real one away. He speaks perfect English, and led the crowd around the town with the rush and energy of a Cook's tourist agent. He took us first through the Cathedral, and showed us in detail things that we could not have seen if we had gone at it alone. Then around to the library and some of the other sights of particular interest, and finally for a spin through the city, to see the damage to the residence district. This was a most interesting beginning, and made a good deal of an impression on our people. They asked questions about the work being done by the people toward cleaning up the ruins of the town and trying to arrange make-shift shelters to live in during the winter. The Mayor is a man of real force of character, and has accomplished marvels under the greatest difficulties.

From Louvain we cut away to the northeast to Aerschot, where we took a quick look at the welter of ruin and struck out to the west through Diest and Haelen, which I saw on my first trip with Frederick Palmer before there was anything done to them.

We got to Liège about one o'clock and had lunch in a restaurant downtown, where we were joined by Jackson, our delegate sent down there to supervise the distribution of food for the Commission. He told us a lot about the difficulties and incidents of his work, and some details of which we had to think. He is the first delegate we have sent to outlying cities, and is up on his toes with interest. A lot more have already sailed from New York, and will soon be here. They are to be spread all over the country in the principal centres, some to stay in the big cities and watch local conditions, and others to travel about their districts and keep track of the needs of the different villages. It is all working out a lot better than we had hoped for, and we have good reason to be pleased. Our chief annoyance is that every time things get into a comfortable state, some idiot starts the story either in England or America that the Germans have begun to seize foodstuffs consigned to us. Then we have to issue statements and get off telegrams, and get renewed assurances from the German authorities and make ourselves a general nuisance to everybody concerned. If we can choke off such idiots, our work will be a lot easier.

The Burgomaster came into the restaurant to find us, and offered to go on with us to Visé, to show us the town, and we were glad to have him, as he knows the place like the palm of his hand.

I had been through Visé twice, and had marvelled at the completeness of the destruction, but had really had no idea of what it was. It was a town of about forty-five hundred souls, built on the side of a pretty hill overlooking the Meuse. There are only two or three houses left. We saw one old man, two children and a cat in the place. Where the others are, nobody knows. The old man was well over sixty, and had that afternoon been put off a train from Germany, where he had been as a prisoner of war since the middle of

August. He had KRIEGSGEFANGENER MUNSTER
stencilled on his coat, front and back, so that there
could be no doubt as to who he was. He was standing
in the street with the tears rolling down his cheeks
and did not know where to go; he had spent the day
wandering about the neighbouring villages trying to
find news of his wife, and had just learned that she had
died a month or more ago. It was getting dark, and
to see this poor old chap standing in the midst of this
welter of ruin without a chick or child or place to lay
his head. . . . It caught our companions hard, and
they loaded the old man up with bank-notes, which
was about all that anybody could do for him and
then we went our way. We wandered through street
after street of ruined houses, sometimes whole blocks
together where there were not enough walls left to
make even temporary shelters.

Near the station we were shown a shallow grave
dug just in front of a house. We were told who filled
the grave—an old chap of over sixty. He had been
made to dig his own grave, and then was tied to a
young tree and shot. The bullets cut the tree in two
just a little above the height of his waist, and the
low wall behind was full of bullet holes.

As nearly as we can learn, the Germans appear to
have come through the town on their way toward
Liège. Nothing was supposed to have happened then,
but on the 15th, 16th and 17th, troops came back from
Liège and systematically reduced the place to ruins
and dispersed the population. It was clear that the
fires were all set, and there were no evidence of street
fighting. It is said that some two hundred civilians

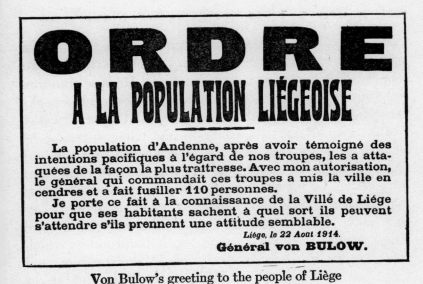

ORDRE
A LA POPULATION LIÉGEOISE

La population d'Andenne, après avoir témoigné des intentions pacifiques à l'égard de nos troupes, les a attaquées de la façon la plus traitresse. Avec mon autorisation, le général qui commandait ces troupes a mis la ville en cendres et a fait fusiller 110 personnes.

Je porte ce fait à la connaissance de la Villé de Liége pour que ses habitants sachent à quel sort ils peuvent s'attendre s'ils prennent une attitude semblable.

Liége, le 22 Aout 1914.
Général von BULOW.

Von Bulow's greeting to the people of Liège

Translation:

ORDER TO THE POPULATION OF LIÈGE

The population of Andenne, after manifesting peaceful intentions toward our troops, attacked them in the most treacherous manner. With my authorization the general who commanded these troops has reduced the town to ashes and has shot 110 persons.

I bring this fact to the knowledge of the City of Liège so that its people may understand the fate which awaits them if they assume a like attitude.

Dieses Haus ist zu Schützen

Es ist streng verboten, ohne Genehmigung der Kommandantur, Haüser zu betreden oder in Brand zu setzen.

Die Etappen-Kommandantur.

How the simple pleasures of the German soldier were restricted

Translation:

THIS HOUSE IS TO BE PROTECTED

It is strictly forbidden to enter houses or set them on fire without the permission of the Kommandantur

were shot, and seven hundred men bundled aboard trains and sent back to Germany as prisoners of war— harmless people like the old chap we saw.

The Burgomaster set out on foot to walk back three kilometers and catch a tram to Liège, and we went southeast to Dalhem, where we spent the night at the Château de Dalhem, on a hill overlooking the picturesque little village snuggled in the bottom of the valley. It was off the main line of march, and had not suffered. The château belongs to General Thyss, who was a great friend of the late King Leopold. He was not there, but the place was being protected by a splendid old dragon in the shape of a German governess who had been with the family for over thirty years, and refused to leave when the war broke out. She had been obliged to lodge a crowd of German officers and some of their men, but held them down with an iron hand, kept them from doing any damage and made them pay for every egg and every bottle of wine they had. We arrived after dark and threw the place into a panic of fear, but Monsieur Francqui soon reassured everybody, and the place was lighted up and placed at our disposal in short order.

Although it was pitch-dark when we arrived, it was only half-past four and we set out on foot to stretch a little. The moon came out and lighted our way through the country roads. We tramped for a couple of hours through all sorts of little towns and villages and groups of houses, some of them wiped out and some hardly touched.

General Thyss's cellars are famous, and with our

dinner of soup and bacon and eggs, we had some of the finest Burgundy I have ever tasted. Early to bed so that we could be up and off at daybreak.

Friday morning we were away early, and made for Herve, where I had never been before. It is a ruin with a few natives and a lot of Landsturm left. We talked to some peasants and to an old priest who gave us something to think about in their stories of happenings there during and after the occupation of their homes. From there to Liège, by way of a lot of little villages whose names I don't remember, but whose condition was pretty bad, past the fort of Fléron and the defensive works that are being put up there.

Wasted some time trying to get gasoline for the other motors, and then the long stretch to Namur, down the valley of the Meuse, and stopped long enough for a look at Andennes, my second visit to the place.

In Andenne and Seilles (a little village across the Meuse) the Germans did a thorough job. They killed about three hundred people and burned about the same number of houses. Most of the houses had been looted systematically. According to the stories of those inhabitants who remain, there was a reign of terror for about a week, during which the Germans rendered themselves guilty of every sort of atrocity and barbarity. They are all most positive that there was no firing upon the German troops by the civil population. It seems to be generally believed that the massacre was due to resistance of retiring Belgian troops and the destruction of bridges and tunnels to cover their retreat. Whatever the provocation, the behaviour of the Germans was that of savages. We were shown photographs showing the corpses of some of those killed. It was to be inferred that they had been wantonly mutilated.

Had lunch at an hotel across the street from the station. After a hasty lunch we made off to Dinant, still following the Meuse. The thin line of houses down the course of the river were thinner than they were a few months ago, and there were signs of suffering and distress everywhere. I had never been to Dinant before, but had seen pictures of it and thought I had an idea of what we were going to see. But the pictures did not give a hint of the horror of the place. The little town, which must have been a gem, nestled at the foot of a huge gray cliff, crowned with the obsolete fort, which was not used or attacked. The town is *gone*. Part of the church is standing, and the walls of a number of buildings, but for the most part, there is nothing but a mess of scattered bricks to show where the houses had stood. And why it was done, we were not able to learn, for everybody there says that there was no fighting in the town itself. We heard stories, too, and such stories that they can hardly be put on paper. Our three guests were more and more impressed as we went on. The bridge was blown up and had fallen into the river, and as we had little time to make the rest of our day's journey, we did not wait to cross by the emergency bridge farther up the river. While we were standing talking to a schoolmaster and his father by the destroyed bridge, seven big huskies with rifles and fixed bayonets came through, leading an old man and a woman who had been found with a camera in their possession. At first there was no objection raised to the taking of photographs, but now our friends are getting a little touchy about it, and lock up anybody silly enough to get caught with kodaks or cameras.

According to what we were told, the Germans entered the town from the direction of Ciney, on the evening of August 21st, and began firing into the windows of the houses. The Germans admit this, but say that there were French troops in the town and this was the only way they could get them out. A few people were killed, but there was nothing that evening in the nature of a general massacre. Although the next day was comparatively quiet, a good part of the population took refuge in the surrounding hills.

On Sunday morning, the 23rd, the German troops set out to pillage and shoot. They drove the people into the street, and set fire to their houses. Those who tried to run away were shot down in their tracks. The congregation was taken from the church, and fifty of the men were shot. All the civilians who could be rounded up were driven into the big square and kept there until evening. About six o'clock the women were lined up on one side of the square and kept in line by soldiers. On the other side, the men were lined up along a wall, in two rows, the first kneeling. Then, under command of an officer, two volleys were fired into them. The dead and wounded were left together until the Germans got round to burying them, when practically all were dead. This was only one of several wholesale executions. The Germans do not seem to contradict the essential facts, but merely

AUX HABITANTS
DE LA
BELGIQUE

Le Maréchal Von der Goltz fait connaître aux Populations de Belgique qu'il est informé par les Généraux Commandants les troupes d'occupation sur le territoire français, que le choléra sévit avec intensité dans les troupes alliées, et qu'il y a le plus grand danger à franchir ces lignes, ou à pénétrer dans le territoire ennemi.

Nous invitons les Populations de Belgique à ne pas entreindre cet avis, et ceux qui croiraient ne pas devoir se soumettre à cet avis, seront traduits devant les Officiers de la Justice Impériale, et nous les prévenons que la peine peut-être celle de mort.

Maréchal Von der Goltz

Septembre 1914

Translation:
Field-Marshal von der Goltz announces to the Belgian population that he is informed by the Generals commanding the troops occupying French territory that cholera is raging fiercely among the allied troops and that there is the greatest danger in crossing the lines or entering enemy territory.

We call upon the Belgian population not to infringe this notice. Those who do not comply with this notice will be brought before the Imperial Officers of Justice and we warn them that the penalty of death may be inflicted upon them.

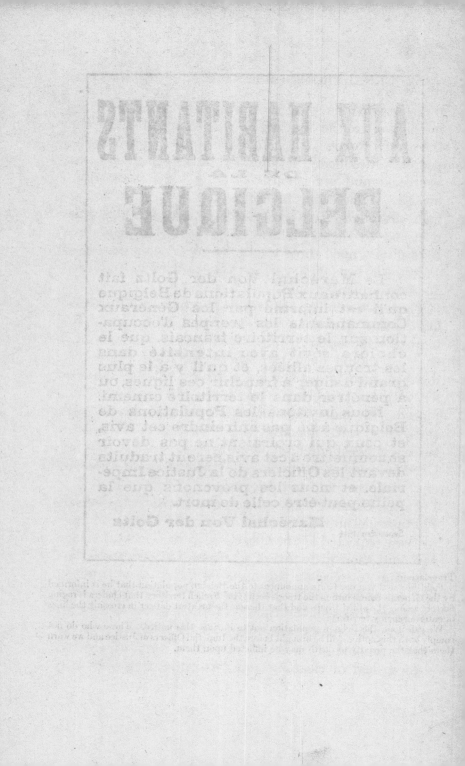

put forward the plea that most of the damage was incidental to the fighting which took place between the armed forces. Altogether more than eight hundred people were killed. Six hundred and twelve have been identified and given burial. Others were not recognisable. I have one of the lists which are still to be had, although the Germans have ordered all copies returned to them. Those killed ranged in age from Félix Fivet, aged three weeks, to an old woman named Jadot, who was eighty. But then Félix probably fired on the German troops.

There is no end to the stories of individual atrocities. One is that Monsieur Wasseige, director of one of the banks, was seized by the Germans, who demanded that he should open the safes. He flatly refused to do this, even under threat of death. Finally he was led with his two eldest sons to the Place d'Armes and placed with more than one hundred others, who were then killed with machine guns. Monsieur Wasseige's three youngest children were brought to the spot by German soldiers, and compelled to witness the murder of their father and two brothers.

From Dinant we struck across country through Phillipeville and some little by-roads to Rance, where we were expected at the house of G. D——. He and his wife and their little girl of five had just returned that morning to receive us, but the place was brightly lighted and as completely prepared as though they had been there all the time. It was a lovely old place, and we were soon made comfortable. German officers have occupied it most of the time, and it required a good deal of cleaning and repairing after they left, but fortunately this work had just been completed, and we had a chance to enjoy the place before any more enforced guests appeared. One of the Imperial princelings had been there for one night, and his name was chalked on the door of his room. He had been *très aimable*, and when he left had taken D——'s motor with him.

We took a tramp around the town in a biting wind, and looked at some of the houses of our neighbours. Some of them were almost wrecked after having served as quarters for troops for varying periods. From others all the furniture had been taken away and shipped back to Germany. One man showed us a card which he had found in the frame of one of his best pictures. It was the card of a German officer, and under the name was written an order to send the picture to a certain address in Berlin. The picture was gone, but the frame and card were still there and are being kept against the day of reckoning—if any. We were shown several little safes which had been pried open and looted, and were told the usual set of stories of what had happened when the army went through. Some of the things would be hard to believe if one did not hear them from the lips of people who are reliable and who live in such widely separated parts of the country at a time when communications are almost impossible.

We had a good and ingeniously arranged dinner. All sorts of ordinary foods are not to be had in this part of the country, and our hostess had, by able thinking, arranged a meal which skillfully concealed the things that were lacking. Among other things, I observed that we had a series of most delicious wines

A Belgian relief ship at Rotterdam

Rotterdam office of the Commission for Relief in Belgium

Barges of the Commission for Relief in Belgium leaving
Rotterdam with cargoes of food

—for our host of that evening also had a wonderful cellar. They had told us just before dinner that the Germans had taken an inventory of their wines and had forbidden them to touch another drop, so I wondered whether they were not incurring some risk in order to give us the wine that they considered indispensable. When I asked our hostess, she told me that it was very simple, that all they needed to do was to drink a part of several bottles, refill them partially with water, seal them, and put them back in the cellars; she said scornfully that "*les Boches* don't know one wine from another," and had not yet been able to detect the fraud. They had a lot of cheap champagne in the cellar and had been filling them up with that, as they prefer any champagne to the best vintage Burgundies. Once in a while there is a little satisfaction reserved for a Belgian.

We were called at daybreak and were on the road at eight o'clock, taking in a series of small villages which had been destroyed, and talking with the few people to be found about the place. This part of Belgium is far worse than the northern part, where the people can get away with comparative ease to one of the larger towns and come back now and then to look after their crops. Here one village after another is wiped out, and the peasants have no place to go unless they travel so far that there is no hope of returning, perhaps for months together. It will be a great problem to provide shelter for these people so that they can return.

We cut through Beaumont, and then took the main road to Mons, where we arrived in the middle of the morning. On the way we had heard that the English nurses had not yet been released, so I made for the military headquarters and saw the commandant. It was evident that they had been hauled over the coals for the way they had behaved when Jack was there, for I never saw such politeness in any headquarters. I was preceded by bowing and unctuous soldiers and non-commissioned officers, all the way from the door to the Presence, and was received by the old man standing. He was most solicitous for my comfort and offered me everything but the freedom of the city. He said that he had not received a word of instructions until a few minutes before my arrival, but that he was now able to give the young ladies their liberty and turn them over to me. In order to get them, I was prayed to go over to the headquarters of the military governor of the Province, and an officer was assigned to accompany me. While we were there, the officer who had been so insulting to Jack and to Miss Hozier came into the room, took one look at us, and scuttled for safety. We heard afterward that he had been ordered to apologise for his behaviour.

At the door of the Provincial headquarters I found another car flying the Legation flag, and Monsieur de Leval came charging out into my arms. There had been a pretty hot time about the nurses and he had finally been sent down to get them out. In a few minutes we had them sitting on a bench in the Governor's office, while Kracker, who used to be one of the Secretaries of the German Legation here, was making out their *laisser-passers* to come to Brussels. They were a happy crowd, but pretty well done up by the treatment they had had.

When they were all fixed I went in and asked for the release of Miss Bradford, another English nurse, who had been in prison in Mons and Charleroi for the past five weeks. I learned of her imprisonment almost by accident while we were waiting for the passports. After some argument it was granted, and I went with a soldier to the prison to get her out. I had not expected to find anything very luxurious, but I was shocked when I saw the place. It was the most severe, repressive penitentiary in the country—still filled with common criminals—and the English nurse was given the same treatment and rations as the worst murderer of the lot. There was the usual row with the man in charge of the place, and finally a soldier was despatched, to tell the young woman she could get ready to go. While she was getting ready, the director of the prison took me around and showed me with great pride things that made me shiver. He said, however, that it was an outrage to put a woman in such a place. The prisoners who do the work of the prison were going about the corridors under guard, each one wearing a dirty brown mask covering his entire head, and with only the smallest of slits for his eyes. They are never allowed to see each other's faces or to speak to one another. I was taken up to the chapel, where each man is herded into a little box like a confessional and locked in so that he cannot see his neighbour, and can only look up toward the raised altar in the centre, where he can see the priest. The school was arranged in the same way, and was shown with equal pride. I fear the jailer thought me lacking in appreciation.

I finally got the young woman out, nearly hysterical, and took her up to the headquarters, and from there to the hotel, where Monsieur de Leval had gathered his charges for luncheon. They were rapidly recovering their old-time spirits, and were chattering away like a lot of magpies.

While I was fussing about with them, I had sent my friends and fellow-travellers ahead, and now left the flock of nurses in the hands of Monsieur de Leval, to be conveyed by tram back to Brussels, while I tried to catch up with my party at the château of Monsieur Warroqué, at Mariemont. I made as much speed as my little car was capable of, but it was nearly two o'clock when I arrived.

The old château of Mariemont is one of those built by Louis XIV, when he set out to have one for each month of the year. This was his place for August. It had been destroyed, and the new one is built near the ruins, but the large park is as it has been for a long time, and a lovely place it is. There were about twenty at table when we arrived, and places were ready for us. More fine wines, and this time to show that we were in the house of a connoisseur, the flunky, in pouring out the precious stuff, would whisper in your ear the name and vintage. Warroqué owns a lot of the coal mines and other properties and is apparently greatly loved by the people. When the Germans came, they seized him as a hostage, but the people became so threatening that he was released. How many men in his position could have counted on that much devotion?

Immediately after luncheon we shoved off and made through the rain for Charleroi, where we took a look at the damage done to the town. It was already dark and we then turned toward Brussels and burned up the road, getting to the Legation at half-past six, to find all the nurses sitting up, having tea with Mrs. Whitlock and the Minister.

Brussels, December 10, 1914.—Yesterday afternoon we received the call of General Freiherr von Bissing, Governor General in Belgium, and of General Freiherr von Kraewel, Military Governor of Brussels. They were accompanied by their suites in full regalia. The military men were most affable, but we did not get any farther than tea and cigarettes. They talked mournfully of the war and said they wished to goodness the whole thing was over. It was a great contrast to the cock-sure talk at the beginning of the war. Von Bissing said that there were hospitals in every village in Germany and that they were all filled with wounded. It is becoming clearer every day that the Germans, as well as others, are getting thoroughly sick and tired of the whole business and would give a lot to end it.

A little while ago the *London Times* cost as high as two hundred francs. It has been going down steadily, until it can be had now for four francs and sometimes for as little as two. The penalties are very severe, but the supply keeps up, although the blockade runners are being picked up every day.

Brussels, December 11, 1914.—This afternoon late B—— brought an uncle to see me, to talk about conditions in France between the Belgian frontier and the German lines. Those poor people cannot, of course, get anything from the heart of France, and as the Belgian frontier is closed tight by the Germans, they are already starving. It looks very much as though we should have to extend the scope of our work, so as to look after them, too. We hear very little news from that part of the country, but from what we do hear, conditions must be frightful. In one little town Mr. K—— came through, only twenty out of five hundred houses are said to be standing. He says that the people are not permitted to leave the place and are living in the cellars and ruins in great misery and practically without food.

Out of a clear sky comes a new trouble for the country. The German Government has come down with a demand for money on a scale that leaves them speechless. The Belgians are ordered to make a forced payment each month of forty millions of francs, for twelve months. The two first payments are to be made by the 15th of next month, and the subsequent installments on the 10th of succeeding months. It is a staggering total, but the German authorities are deaf to appeals, and the Provinces will have to get together and raise the money in some way.

An entry from a later part of Mr. Gibson's journal gives a picture of the Belgian spirit under German rule and one of the few methods of retaliation they had against German oppression.

The Belgians are getting a good deal of quiet pleasure these days from a clandestine newspaper called *La Libre Belgique* which is published almost in the shadow of the

Brussels, December 14, 1914.—Yesterday afternoon late, after a session at C. R. B.* headquarters, I dropped in for a cup of tea with Baronne Q——. There was a fine circle of gossip and I learned all the spicy stuff. The husband of Mme. de F—— had been in prison for a month, having been pulled out of a motor on his way to the frontier, and found with letters on him. He got out on Thursday and they are quite proud of them-

Kommandantur. It is a little four-page paper that is published "every now and then" and says anything it likes about the "Occupant." It also publishes news and texts that are barred from the censored press. It is distributed in a mysterious way that still has the Germans guessing, although they have detailed their cleverest sleuths to the task of hunting down the paper and those responsible for its publication. Every number is delivered to all the more important German officials in Brussels and, more remarkable still, it appears without fail upon the desk of the Governor-General—in that sanctum guarded like the vaults of the Bank of England. Sometimes it appears in the letter-box in the guise of a letter from Germany; sometimes it is thrown in the window; sometimes it is delivered by an orderly with a bundle of official despatches; sometimes it merely appears from nowhere. But it never fails to reach the Governor-General. He never fails to read it and to wax wroth over its contents. Large rewards have been offered for information about the people who are writing and printing the paper. The Germans rage publicly, which only adds to the pleasure that the Belgians get from their little enterprise.

My copy reaches me regularly and always in some weird way as in the case of the Germans. I don't know who my friend is that sends me the paper. Whoever he is I am much obliged.

* Commission for Relief in Belgium. This name was given the original American Relief Committee within a few weeks of its foundation.

selves. They were having a fine time discussing the predicament of the H—— family. The Countess was arrested last week because she, too, was caught carrying letters. She was released from prison and allowed to return home. Now the Germans have placed sentries before the house and allow no one to enter or leave. The old gentleman is also locked up there. The servants have been driven out, and are not even permitted to bring meals to their *patrons*, who are dependent on what they are given to eat by the German soldiers. There is no charge against them at present, so they have no idea as to how long the present charming situation will last. There was a great amount of gossip and the right amount of tea and cakes, so I had an enjoyable half hour.

Yesterday morning Grant-Watson was put aboard a train and taken to Berlin, where he is to be guarded as a prisoner of war. It is all most outrageous, as Lancken definitely promised that he would not be molested. Moral: get just as far away from these people as you can, while you can, in the knowledge that if they "change their mind," promises won't count.

Jeffes is left here for the present and may be released. We shall try to get him off, but in view of what has already happened, cannot be very confident. Jeffes is philosophical and uncomplaining, but naturally is not very happy.

Brussels, Sunday, December 20, 1914.—Jack got off to London yesterday after a visit of six weeks. Had it not been for the nearness of Christmas and the knowledge that he was needed at home, he would have

NUMÉRO 30   JUIN 1915

PRIX DU NUMÉRO — élastique, de zéro à l'infini (prière aux revendeurs de ne pas dépasser cette limite)

LA LIBRE BELGIQUE

Acceptons provisoirement les sacrifices qui nous sont imposés........
et attendons patiemment l'heure de la réparation.

Le Bourgmestre
ADOLPHE MAX.

FONDÉE
LE 1er FÉVRIER 1915

Envers ces personnes qui dominent par la force militaire notre pays,
ayons les égards que commande l'intérêt général. Respectons les
règlements qu'elles nous imposent aussi longtemps qu'ils ne portent
atteinte ni à la liberté de nos consciences chrétiennes ni à notre
Dignité Patriotique.
Mgr MERCIER.

BULLETIN DE PROPAGANDE PATRIOTIQUE — RÉGULIÈREMENT IRRÉGULIER
NE SE SOUMETTANT A AUCUNE CENSURE

ADRESSE TÉLÉGRAPHIQUE	BUREAUX ET ADMINISTRATION	ANNONCES : Les affaires étant nulles
KOMMANDANTUR - BRUXELLES	ne pouvant être un emplacement de tout repos, ils sont installés dans une cave automobile	sous la domination allemande, nous avons supprimé la page d'annonces et conseillons à nos clients de réserver leur argent pour des temps meilleurs.

AVIS.

On nous fait à nouveau l'honneur de s'occuper de notre modeste bulletin. Nous en sommes flattés, mais nous nous voyons, forcés de répéter ce que nous avons déjà dit pour notre défense. Ce n'est certes pas nous qu'on peut accuser sans manquer à la vérité, de provoquer nos concitoyens à la révolte. Nous ne manquons pas une occasion de prêcher la patience, l'endurance, le calme et le respect des lois de la guerre. Aussi profitons-nous de cette occasion qui nous est offerte pour répéter l'avis que nous avons déjà inséré :

RESTONS CALMES!!!

Le jour viendra (lentement mais sûrement) ou nos ennemis contraints de reculer devant les Alliés, devront abandonner notre capitale.

Souvenons-nous alors des avis nombreux qui ont été donnés aux civils par le Gouvernement et par notre bourgmestre

SON EXCELLENCE LE GOUVERNEUR Bon VON BISSING ET SON AMIE INTIME

NOTRE CHER GOUVERNEUR, ÉCŒURÉ PAR LA LECTURE
DES MENSONGES DES JOURNAUX CENSURÉS, CHERCHE LA VÉRITÉ
DANS LA « LIBRE BELGIQUE »

M. Max : SOYONS CALMES!!!

Faisons taire les sentiments de légitime colère qui fermentent en nos cœurs.

Soyons, comme nous l'avons été jusqu'ici, respectueux des lois de la guerre. C'est ainsi que nous continuerons à mériter l'estime et l'admiration de tous les peuples civilisés.

Ce serait une INUTILE LACHETÉ*, une lâcheté indigne des Belges que chercher à se venger ailleurs que sur le champ de bataille. Ce serait de plus* EXPOSER DES INNOCENTS *à des représailles terribles de la part d'ennemis sans pitié et sans justice.*

Méfions-nous des agents provocateurs allemands qui, en exaltant notre patriotisme, nous pousseraient à commettre des excès.

RESTONS MAITRES DE NOUS-MÊMES ET PRÊCHONS LE CALME AUTOUR DE NOUS. C'EST LE PLUS GRAND SERVICE QUE NOUS PUISSIONS RENDRE A NOTRE CHÈRE PATRIE.

L'ORDRE SOCIAL TOUT ENTIER DEFENDU PAR LA BELGIQUE.

Le 3 août, le Gouvernement allemand remet à la Belgique une note demandant le libre passage pour ses armées sur son territoire, moyennant quoi l'Allemagne s'engage à maintenir l'intégrité du royaume et de ses possessions, Sinon, la Belgique sera traitée en ennemie. Le roi Albert a douze heures pour répondre. Devant cet ultimatum, il n'hésite pas. Il sait que l'armée allemande est une force terrible. Il connaît l'empe-

reur allemand. Il sait que l'orgueilleux, après une telle démarche, ne reculera plus. Son trône est en jeu, plus que son trône . les sept millions d'âmes — quelle éloquence prennent les vulgaires termes des statistiques dans certaines circonstances ! — qui lui sont confiées . il voit en esprit ce beau pays indéfendable ces charbonnages, ces carrières, ces usines, ces filatures, ces ports, cette florissante industrie épanouie dans ces plaines ouvertes qu'il ne pourra pas préserver. Mais il s'agit d'un traité où il y a sa signature. Répondre oui à l'Allemagne, c'est trahir ses consignataires, le

PRIÈRE DE FAIRE CIRCULER CE BULLETIN

La Libre Belgique, the clandestine paper printed in Brussels in 1915, which survived General von Bissing's reward for the discovery of its office and made fun of him by faking a picture of him reading their condemned paper.

Edward D. Curtis. The first volunteer worker of the Commission for Relief in Belgium. He served continuously from the autumn of 1914 until the entry of the United States into the war

I have learned with gratification of the noble and effective work being done by American citizens and officials on behalf of my stricken people. I confidently hope that their efforts will receive that ungrudging support which we have learned to expect from the generous womanhood of America.

We mothers of Belgium no less than the mothers of America have for generations instilled in our children the instincts and the love of peace. We asked no greater boon than to live in peace and friendship with all the world. We have provoked no war, yet in defense of our hearthstones, our country has been laid waste from end to end.

The flow of commerce has ceased and my people are faced with famine. The terrors of starvation with its consequences of disease and violence menace the unoffending civilian population — the aged, the infirm, the women and the children.

American officials and citizens in Belgium and England, alive to their country's traditions, have created an organization under the protection of their Government and are already sending food to my people. I hope that they may receive the fullest sympathy and aid from every side.

I need not say that I and my people shall always hold in grateful remembrance the proven friendship of America in this hour of need.

Elisabeth

Appeal of the Queen of the Belgians for help from America

been prepared to stay on indefinitely. His grief at leaving was genuine. He invested heavily in flowers and chocolates for the people who had been nice to him, endowed all the servants, and left amid the cheers and sobs of the populace. He is a good sort, and I was sorry to see him go. By this time he is probably sitting up in London, telling them all about it.

To-day I went up to Antwerp to bring back our old motor. Left a little before noon, after tidying up my desk, and took my two Spanish colleagues, San Esteban and Molina, along for company. I had the passes and away we went by way of Malines, arriving in time for a late lunch.

Antwerp is completely Germanised already. We heard hardly a word of French anywhere—even the hotel waiters speaking only hotel French. The crowd in the restaurant of the Webber was exclusively German, and there was not a word of French on the menu.

The Germans took over the garage where our car was left the day they came in, and there I discovered what was left of the old machine. The sentries on guard at the door reluctantly let us in, and the poor proprietor of the garage led us to the place where our car has stood since the fall of Antwerp. The soldiers have removed two of the tires, the lamps, cushions, extra wheels, speedometer, tail lights, tool box, and had smashed most of the other fixings they could not take off. In view of the fact that my return trip to Brussels at the time of the bombardment was for the purpose of bringing the plans of the city to the Germans, so that they would have knowledge of the location of the public monuments and could spare them, it seems rather rough that they should repay us by smashing our motor. I think we shall make some remarks to them to this effect to-morrow, and intimate that it is up to them to have the car repaired and returned to us in good shape.

The first group of Americans to work on the relief came into Belgium this month. They are, for the most part, Rhodes Scholars who were at Oxford, and responded instantly to Hoover's appeal. They are a picked crew, and have gone into the work with enthusiasm. And it takes a lot of enthusiasm to get through the sort of pioneer work they have to do. They have none of the thrill of the fellows who have gone into the flying corps or the ambulance service. They have ahead of them a long winter of motoring about the country in all sorts of weather, wrangling with millers and stevedores, checking cargoes and costs, keeping the peace between the Belgians and the German authorities, observing the rules of the game toward everybody concerned, and above all, keeping neutral. It is no small undertaking for a lot of youngsters hardly out of college, but so far they have done splendidly.

The one I see the most of is Edward Curtis, who sails back and forth to Holland as courier of the Commission. He was at Cambridge when the war broke out, and after working on Hoover's London Committee to help stranded Americans get home, he came on over here and fell to. He exudes silence and discretion, but does not miss any fun or any chance to advance the general cause. Of course it is taking the Germans some time to learn his system. He is absolutely square with them, and gets a certain amount of fun out of

Julius Van Hee, American Vice-Consul at Ghent   Lewis Richards

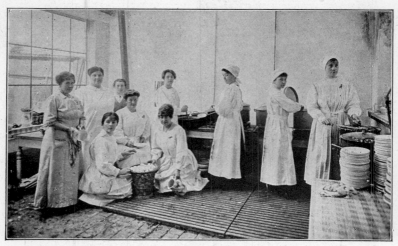

A Brussels soup-kitchen run by volunteers

Meals served to the children in the schools

their determined efforts to find some sort of contra-
band on him. They can hardly conceive of his being
honest, and think his seeming frankness is merely an
unusually clever dodge to cover up his transgressions.

Brussels, December 21, 1914.—Yesterday Brussels
awoke from the calm in which it had been plunged for
some time, when a couple of French aviators came
sailing overhead and dropped six bombs on the railroad yards at Etterbeck. I was away at Antwerp
and did not see it, but everybody else of the population
of 700,000 Bruxellois did, and each one of them has
given me a detailed account of it. The German forces
did their level best to bring the bird men down with
shrapnel, but they were flying high enough for safety.
They seem to have hit their mark and torn up the
switches, etc., in a very satisfactory way. For three
or four days we have been hearing the big guns again,
each day more distinctly; but we don't know what
it means. The Germans explain it on the ground that
they are testing guns.

Mr. and Mrs. Hoover arrived last night, bringing
Frederick Palmer with them. We dined together at
the Palace. They were full of news, both war and
shop, and I sat and talked with them until after eleven,
greatly to the prejudice of my work. Had to stay up
and grind until nearly two.

Curtis, who came back last night, says that Jack
was arrested at Antwerp on his way out, because he
had Folkstone labels on his bags. It took him so long
to explain away his suspicious belongings that he barely
caught the last train from Rosendaal to Flushing. He

seems to be destined to a certain amount of arrest now and then.

Hoover turned up at the Legation this morning at a little after nine, and he and the Minister and I talked steadily for three hours and a half.

Despite the roar of work at the Legation, I went off after lunch with Mrs. Whitlock and did some Xmas shopping—ordered some flowers and chocolates. Went out and dropped Mrs. Whitlock at Mrs. B——'s, to help decorate the tree she is going to have for the English children here. B—— is a prisoner at Ruhleben, and will probably be there indefinitely, but his wife is a trump. She had a cheery letter from him, saying that he and his companions in misery had organised a theatrical troupe, and were going soon to produce *The Importance of Being Earnest*.

Brussels, Christmas, 1914.—This is the weirdest Christmas that ever was—with no one so much as thinking of saying "Merry Christmas." Everything is so completely overshadowed by the war, that had it not been for the children, we should have let it go unnoticed.

Yesterday evening there was a dinner at the Legation—Bicknell, Rose and James, the Hoovers and Frederick Palmer. Although there was a bunch of mistletoe over the table, it did not seem a bit Christmasy, but just an ordinary good dinner with much interesting talk.

Immediately after lunch we climbed into the big car and went out to Lewis Richards' Christmas tree. He has a big house at the edge of town, with grounds which were fairy-like in the heavy white frost. He had undertaken to look after 660 children, and he did it to the Queen's taste. They were brought in by their mothers in bunches of one hundred, and marched around the house, collecting things as they went. In one room each youngster was given a complete outfit of warm clothes. In another, some sort of a toy which he was allowed to choose. In another, a big bag of cakes and candies, and, finally, they were herded into the big dining-room, where they were filled with all sorts of Xmas food. There was a big tree in the hall, so that the children, in their triumphal progress, merely walked around the tree. Stevens had painted all the figures and the background of an exquisite *crèche*, with an electric light behind it, to make the stars shine. The children were speechless with happiness, and many of the mothers were crying as they came by.

Since the question of food for children became acute here, Richards has been supplying rations to the babies in his neighbourhood. The number has been steadily increasing, and for some time he has been feeding over two hundred youngsters a day. He has been very quiet about it, and hardly anyone has known what he was doing.

It is cheering to see a man who does so much to comfort others; not so much because he weighs the responsibility of his position and fortune, but because he has a great-hearted sympathy and instinctively reaches out to help those in distress. Otherwise the day was pretty black, but it did warm the cockles of my heart to find this simple American putting some real meaning into Christmas for these hundreds of wretched people. He also gave it a deeper meaning for the rest of us.

Brussels, December 31, 1914—Here is the end of the vile old year. We could see it out with rejoicing, if there were any prospect of 1915 bringing us anything better. But it doesn't look very bright for Belgium.

THE CASE OF MISS EDITH CAVELL

*The extracts from this journal have been so volu-
minous as to preclude bringing the record much farther
than the end of 1914. In the main the story of 1915–
1916 is in the development of the Commission for Relief
in Belgium and the new light shed each day upon
German methods and mentality. It is a long story
and could not be crowded between the covers of this
volume. There is, however, one outstanding event in
1915—the case of Miss Edith Cavell—which is of
such interest and so enlightening as to conditions in
Belgium under German domination as to warrant its
inclusion in this book. At the risk, therefore, of appear-
ing disconnected it has been decided to publish as a
final chapter an article in regard to the case of Miss
Cavell which has already appeared in the "World's
Work."*

On August 5, 1915, Miss Edith Cavell, an English-
woman, directress of a large nursing home at Brussels,
was quietly arrested by the German authorities and
confined in the prison of St. Gilles on the charge that
she had aided stragglers from the Allied armies to
escape across the frontier from Belgium to Holland,
furnishing them with money, clothing and information
concerning the route to be followed. It was some

time before news of Miss Cavell's arrest was received by the American Legation, which was entrusted with the protection of British interests in the occupied portion of Belgium. When the Minister at Brussels received a communication from the Ambassador at London transmitting a note from the Foreign Office stating that Miss Cavell was reported to have been arrested and asking that steps be taken to render her assistance, Mr. Whitlock immediately addressed a note to the German authorities asking whether there was any truth in the report of Miss Cavell's arrest and requesting authorisation for Maître Gaston de Leval, the legal counselor of the Legation, to consult with Miss Cavell and, if desirable, entrust some one with her defense.

No reply was received to this communication, and on September 10th the Legation addressed a further note to Baron von der Lancken, Chief of the Political Department, calling his attention to the matter and asking that he enable the Legation to take such steps as might be necessary for Miss Cavell's defense.

On September 12th a reply was received from Baron von der Lancken in which it was stated that Miss Cavell had been arrested on August 5th and was still in the military prison of St. Gilles. The note continued:

She has herself admitted that she concealed in her house French and English soldiers, as well as Belgians of military age, all desirous of proceeding to the front. She has also admitted having furnished these soldiers with the money necessary for their journey to France, and having facilitated their departure from Belgium by providing them with guides, who enabled them to cross the Dutch frontier secretly.

Miss Cavell's defense is in the hands of the advocate Braun, who, I may add, is already in touch with the competent German authorities.

In view of the fact that the Department of the Governor-General, as a matter of principle, does not allow accused persons to have any interviews whatever, I much regret my inability to procure for M. de Leval permission to visit Miss Cavell as long as she is in solitary confinement.

Under the provisions of international law the American Minister could take no action while the case was before the courts. It is an elementary rule that the forms of a trial must be gone through without interference from any source. If, when the sentence has been rendered, it appears that there has been a denial of justice, the case may be taken up diplomatically, with a view to securing real justice. Thus in the early stages of the case the American Minister was helpless to interfere. All that he could do while the case was before the courts was to watch the procedure carefully and be prepared with a full knowledge of the facts to see that a fair trial was granted.

Maître de Leval communicated with Mr. Braun, who said that he had been prevented from pleading before the court on behalf of Miss Cavell, but had asked his friend and colleague, Mr. Kirschen, to take up the case. Maître de Leval then communicated with Mr. Kirschen, and learned from him that lawyers defending prisoners before German military courts were not allowed to see their clients before the trial and were shown none of the documents of the prosecution. It was thus manifestly impossible to prepare any defense save in the presence of the court and during the progress of the trial. Maître de Leval, who

PROCLAMATION

Le Tribunal du Conseil de Guerre Impérial Allemand siégeant à Bruxelles a prononcé les condamnations suivantes :

Sont condamnés à mort pour trahison en bande organisée :

Edith CAVELL, Institutrice à Bruxelles.
Philippe BANCQ, Architecte à Bruxelles.
Jeanne de BELLEVILLE, de Montignies.
Louise THUILIEZ, Professeur a Lille.
Louis SEVERIN Pharmacien à Bruxelles.
Albert LIBIEZ, Avocat à Mons.

Pour le même motif, ont été condamnés à quinze ans de travaux forcés :

Hermann CAPIAU, Ingénieur à Wasmes. – Ada BODART, à Bruxelles. – Georges DERVEAU, Pharmacien à Pâturages. – Mary de CROY, à Bellignies.

Dans sa même séance, le Conseil de Guerre a prononcé contre dix-sept autres accusés de trahison envers les Armées Impériales, des condamnations de travaux forcés et de prison variant entre deux ans et huit ans.

En ce qui concerne BANCQ et Edith CAVELL, le jugement a déjà reçu pleine exécution.

Le Général Gouverneur de Bruxelles porte ces faits à la connaissance du public pour qu'ils servent d'avertissement.

Bruxelles le 12 Octobre 1915

Le Gouverneur de la Ville,
Général VON BISSING

German proclamation announcing the execution of Miss Cavell

Translation:

The Imperial German Court Martial sitting at Brussels has pronounced the following sentence:

Condemned to death for treason committed as an organized band:

Edith Cavell, teacher, of Brussels.
Philippe Bancq, Architect, of Brussels.
Jeanne de Belleville, of Montignies.
Louise Thuilier, Teacher, of Lille.
Louis Severin, druggist, of Brussels.
Albert Libiez, lawyer, of Mons.

For the same offense the following are condemned to fifteen years of hard labor:

Hermann Capiau, engineer, of Wasmes—Ada Bodart, of Brussels—Georges Derveau, druggist, of Paturages—Mary de Croy, of Bellignies.

At the same session the Court Martial has pronounced sentences of hard labor and of imprisonment, varying from two to eight years, against seventeen others accused of treason against the Imperial Armies.

As regards Bancq and Edith Cavell, the sentence has already been fully carried out.

The Governor-General brings these facts to the attention of the public in order that they may serve as a warning.

from the beginning to the end of the case showed a most serious and chivalrous concern for the welfare of the accused, then told Mr. Kirschen that he would endeavour to be present at the trial in order to watch the case. Mr. Kirschen dissuaded him from attending the trial on the ground that it would only serve to harm Miss Cavell rather than help her; that the judges would resent the presence of a representative of the American Legation. Although it seems unbelievable that any man of judicial mind would resent the presence of another bent solely on watching the course of justice, Mr. Kirschen's advice was confirmed by other Belgian lawyers who had defended prisoners before the German military courts and spoke with the authority of experience. Mr. Kirschen promised, however, to keep Maître de Leval fully posted as to all the developments of the case and the facts brought out in the course of the trial.

The trial began on Thursday, October 7th, and ended the following day.

On Sunday afternoon the Legation learned from persons who had been present at the trial some of the facts.

It seems that Miss Cavell was prosecuted for having helped English and French soldiers, as well as Belgian young men, to cross the frontier into Holland in order that they might get over to England. She had made a signed statement admitting the truth of these charges and had further made public acknowledgment in court. She frankly admitted that not only had she helped the soldiers to cross the frontier but that some of them had written her from England thanking her for her assistance. This last admission made the case more serious for her because if it had been proven only that she had helped men to cross the frontier into Holland, she could have been sentenced only for a violation of the passport regulations, and not for the "crime" of assisting soldiers to reach a country at war with Germany.

Miss Cavell was tried under Paragraph 58 of the German Military Code, which says:

Any person who, with the intention of aiding the hostile Power or causing harm to German or allied troops, is guilty of one of the crimes of Paragraph 90 of the German Penal Code, will be sentenced to death for treason.

The "crime" referred to by Paragraph 90 was that of "conducting soldiers to the enemy" (viz.: *dem Feinde Mannschaften zuführt*).

It is manifest that this was a strained reading of the provisions of military law; that a false interpretation was wilfully put upon these provisions in order to secure a conviction. This law was obviously framed to cover the case of those who assist stragglers or lost soldiers to get back to their own lines and join their units. It is doubtful whether the framers of the military law had foreseen anything so indirect and unprecedented as that of helping soldiers cross into a neutral country in the hope that they might find their way back through two other countries to their own army. Miss Cavell assisted these soldiers to escape into a neutral country which was bound, if possible, to apprehend and intern them. If these soldiers succeeded in outwitting the Dutch authorities and making their way to England, their success would not, to any fair-minded person, increase the offense committed by Miss Cavell.

Miss Cavell's conduct before the court was marked by the greatest frankness and courage. She stated that she had assisted these men to escape into Holland because she thought that if she had not done so they would have been seized and shot by the Germans; that she felt that she had only done her duty in helping to save their lives.

The Military Prosecutor replied that while this argument might be made concerning English soldiers, it could not apply to Belgians, who were free to remain in the country without danger. The subsequent behaviour of the German authorities to the Belgian young men who remained in the country does not lend any considerable weight to the remarks of the Public Prosecutor.

In concluding his plea, the Public Prosecutor asked that the court pass the sentence of death upon Miss Cavell and eight other prisoners among the thirty-five brought to trial.

Upon ascertaining these facts Maître de Leval called at the Political Department and asked that, the trial having taken place, permission be granted him to see Miss Cavell in person, as there could be no further objection to consultation. Herr Conrad, an official of the Political Department, who received Maître de Leval, stated that he would make enquiry of the court and communicate with him later.

The foregoing are the developments up to Sunday night, October 10th. Subsequent developments are shown by the following extracts from a journal made at the time:

Brussels, October 12, 1915.—When I came in yesterday morning I found information which seemed to confirm previous reports that Miss Cavell's trial had been concluded on Saturday afternoon and that the prosecution had asked that the death sentence be imposed. Monsieur de Leval promptly called the Political Department over the telephone and talked to Conrad, repeating our previous requests that he be authorised to see Miss Cavell in prison. He also asked that Mr. Gahan, the English chaplain, be permitted to visit her. Conrad replied that it had been decided that Mr. Gahan could not see her, but that she could see any of the three Protestant clergymen (Germans) attached to the prison; that de Leval could not see her until the judgment was pronounced and signed. He said that as yet no sentence had been pronounced and that there would probably be a delay of a day or two before a decision was reached. He stated that even if the judgment of the court had been given, it would have no effect until it had been confirmed by the Governor, who was absent from Brussels and would not return for two or possibly three days. We asked Conrad to inform the Legation immediately upon the confirmation of the sentence in order that steps might be taken to secure a pardon if the judgment really proved to be one of capital punishment. Conrad said he had no information to the effect that the court had acceded to the request for the death sentence, but promised to keep us informed. I stood by the telephone and could overhear both de Leval and Conrad.

Despite the promise of the German authorities to

keep us fully posted, we were nervous and apprehensive and remained at the Legation all day, making repeated enquiry by telephone to learn whether a decision had been reached. On each of these occasions the Political Department renewed the assurance that we would be informed as soon as there was any news. In order to be prepared for every eventuality, we drew up a petition for clemency addressed to the Governor-General, and a covering note addressed to Baron von der Lancken, in order that they might be presented without loss of time in case of urgent need.

A number of people had been arrested and tried for helping men to cross into Holland, but, so far as we know, the death sentence had never been inflicted. The usual thing was to give a sentence of imprisonment in Germany. The officials at the Political Department professed to be skeptical as to the reported intention of the court to inflict the death sentence, and led us to think that nothing of the sort need be apprehended.

None the less we were haunted by a feeling of impending horror that we could not shake off. I had planned to ride in the afternoon, but when my horse was brought around, I had it sent away and stayed near the telephone. Late in the afternoon de Leval succeeded in getting into communication with a lawyer interested in one of the accused. He said that the German Kommandantur had informed him that judgment would be passed the next morning, Tuesday. He was worried as to what was in store for the prisoners and said he feared the court would be very severe.

At 6.20 I had Topping (clerk of the Legation) telephone Conrad again. Once more we had the most definite assurances that nothing had happened and a somewhat weary renewal of the promise that we should have immediate information when sentence was pronounced.*

At 8.30 I had just gone home when de Leval came for me in my car, saying that he had come to report that Miss Cavell was to be shot during the night. We could hardly credit this, but as our informant was so positive and insisted so earnestly, we set off to see what could be done.

De Leval had seen the Minister, who was ill in bed, and brought me his instructions to find von der Lancken, present the appeal for clemency, and press for a favourable decision. In order to add weight to our representations, I was to seek out the Spanish Minister to get him to go with us and join in our appeal. I found him dining at Baron Lambert's, and on explaining the case to him he willingly agreed to come.

When we got to the Political Department we found that Baron von der Lancken and all the members of his staff had gone out to spend the evening at one of the disreputable little theatres that have sprung up here for the entertainment of the Germans. At first we were unable to find where he had gone, as the orderly on duty evidently had orders not to tell, but by dint of some blustering and impressing on him the fact that Lancken would have cause to regret not having seen us, he agreed to have him notified. We put the

* This was just one hour and twenty minutes after the sentence had actually been pronounced. There is no need for comment.

orderly into the motor and sent him off. The Marquis de Villalobar, de Leval, and I settled down to wait, and we waited long, for Lancken, evidently knowing the purpose of our visit, declined to budge until the end of an act that seemed to appeal to him particularly.

He came in about 10.30, followed shortly by Count Harrach and Baron von Falkenhausen, members of his staff. I briefly explained to him the situation as we understood it and presented the note from the Minister, transmitting the appeal for clemency. Lancken read the note aloud in our presence, showing no feeling aside from cynical annoyance at something—probably our having discovered the intentions of the German authorities.

When he had finished reading the note, Lancken said that he knew nothing of the case, but was sure in any event that no sentence would be executed so soon as we had said. He manifested some surprise, not to say annoyance, that we should give credence to any report in regard to the case which did not come from his Department, that being the only official channel. Leval and I insisted, however, that we had reason to believe our reports were correct and urged him to make inquiries. He then tried to find out the exact source of our information, and became painfully insistent. I did not propose, however, to enlighten him on this point and said that I did not feel at liberty to divulge our source of information.

Lancken then became persuasive—said that it was most improbable that any sentence had been pronounced; that even if it had, it could not be put into effect within so short a time, and that in any event all

Government offices were closed and that it was impossible for him to take any action before morning. He suggested that we all go home "reasonably," sleep quietly, and come back in the morning to talk about the case. It was very clear that if the facts were as we believed them to be, the next morning would be too late, and we pressed for immediate enquiry. I had to be rather insistent on this point, and de Leval, in his anxiety, became so emphatic that I feared he might bring down the wrath of the Germans on his own head, and tried to quiet him. There was something splendid about the way de Leval, a Belgian with nothing to gain and everything to lose, stood up for what he believed to be right and chivalrous, regardless of consequences to himself.

Finally, Lancken agreed to enquire as to the facts, telephoned from his office to the presiding judge of the court martial, and returned in a short time to say that sentence had indeed been passed and that Miss Cavell was to be shot during the night.

We then presented with all the earnestness at our command, the plea for clemency. We pointed out to Lancken that Miss Cavell's offenses were a matter of the past; that she had been in prison for some weeks, thus effectually ending her power for harm; that there was nothing to be gained by shooting her, and on the contrary this would do Germany much more harm than good and England much more good than harm. We pointed out to him that the whole case was a very bad one from Germany's point of view; that the sentence of death had heretofore been imposed only for cases of espionage and that Miss Cavell was not even accused

Miss Edith Cavell

Fly-leaf of Miss Cavell's prayer book

by the German authorities of anything so serious.* We reminded him that Miss Cavell, as directress of a large nursing home, had, since the beginning of the war, cared for large numbers of German soldiers in a way that should make her life sacred to them. I further called his attention to the manifest failure of the Political Department to comply with its repeated promises to keep us informed as to the progress of the trial and the passing of the sentence. The deliberate policy of subterfuge and prevarication by which they had sought to deceive us, as to the progress of the case, was so raw as to require little comment. We all pointed out to Lancken the horror of shooting a woman, no matter what her offense, and endeavoured to impress upon him the frightful effect that such an execution would have throughout the civilised world. With an ill-concealed sneer he replied that on the contrary he was confident that the effect would be excellent.

When everything else had failed, we asked Lancken to look at the case from the point of view solely of German interests, assuring him that the execution of Miss Cavell would do Germany infinite harm. We reminded him of the burning of Louvain and the sinking of the *Lusitania*, and told him that this murder would rank with those two affairs and would stir all civilised countries with horror and disgust. Count

---

* At the time there was no intimation that Miss Cavell was guilty of espionage. It was only when public opinion had been aroused by her execution that the German Government began to refer to her as "the spy Cavell." According to the German statement of the case, there is no possible ground for calling her a spy.

Harrach broke in at this with the rather irrelevant remark that he would rather see Miss Cavell shot than have harm come to the humblest German soldier, and his only regret was that they had not "three or four old English women to shoot."

The Spanish Minister and I tried to prevail upon Lancken to call Great Headquarters at Charleville on the telephone and have the case laid before the Emperor for his decision. Lancken stiffened perceptibly at this suggestion and refused, frankly, saying that he could not do anything of the sort. Turning to Villalobar, he said, "I can't do that sort of thing. I am not a friend of my Sovereign as you are of yours," to which a rejoinder was made that in order to be a good friend, one must be loyal and ready to incur displeasure in case of need. However, our arguments along this line came to nothing, but Lancken finally came to the point of saying that the Military Governor of Brussels was the supreme authority (*Gerichtsherr*) in matters of this sort and that even the Governor-General had no power to intervene. After further argument he agreed to get General von Sauberschweig, the Military Governor, out of bed to learn whether he had already ratified the sentence and whether there was any chance for clemency.

Lancken was gone about half an hour, during which time the three of us laboured with Harrach and Falkenhausen, without, I am sorry to say, the slightest success. When Lancken returned he reported that the Military Governor said that he had acted in this case only after mature deliberation; that the circumstances of Miss Cavell's offense were of such character that he considered infliction of the death penalty imperative. Lancken further explained that under the provisions of German Military Law, the *Gerichtsherr* had discretionary power to accept or to refuse to accept an appeal for clemency; that in this case the Governor regretted that he must decline to accept the appeal for clemency or any representations in regard to the matter.

We then brought up again the question of having the Emperor called on the telephone, but Lancken replied very definitely that the matter had gone too far; that the sentence had been ratified by the Military Governor, and that when matters had gone that far, "even the Emperor himself could not intervene."*

He then asked me to take back the note I had presented to him. I at first demurred, pointing out that this was not an appeal for clemency, but merely a note to him, transmitting a note to the Governor, which was itself to be considered the appeal for clemency. I pointed out that this was especially stated in the Minister's note to him, and tried to prevail upon him to keep it. He was very insistent, however, and inasmuch as he had already read the note aloud to us and we knew that he was aware of its contents, it seemed that there was nothing to be gained by refusing to accept the note, and I accordingly took it back.

Despite Lancken's very positive statements as to the futility of our errand, we continued to appeal to every sentiment to secure delay and time for reconsideration

---

* Although accepted at the time as true, this statement was later found to be entirely false and is understood to have displeased the Emperor. The Emperor could have stopped the execution at any moment.

of the case. The Spanish Minister led Lancken aside and said some things to him that he would have hesitated to say in the presence of Harrach, Falkenhausen, and de Leval, a Belgian subject. Lancken squirmed and blustered by turns, but stuck to his refusal. In the meantime I went after Harrach and Falkenhausen again. This time, throwing modesty to the winds, I reminded them of some of the things we had done for German interests at the outbreak of the war; how we had repatriated thousands of German subjects and cared for their interests; how during the siege of Antwerp I had repeatedly crossed the lines during actual fighting at the request of Field Marshal von der Goltz to look after German interests; how all this service had been rendered gladly and without thought of reward; that since the beginning of the war we had never asked a favour of the German authorities and it seemed incredible that they should now decline to grant us even a day's delay to discuss the case of a poor woman who was, by her imprisonment, prevented from doing further harm, and whose execution in the middle of the night, at the conclusion of a course of trickery and deception, was nothing short of an affront to civilisation. Even when I was ready to abandon all hope, de Leval was unable to believe that the German authorities would persist in their decision, and appealed most touchingly and feelingly to the sense of pity for which we looked in vain.

Our efforts were perfectly useless, however, as the three men with whom we had to deal were so completely callous and indifferent that they were in no way moved by anything that we could say.

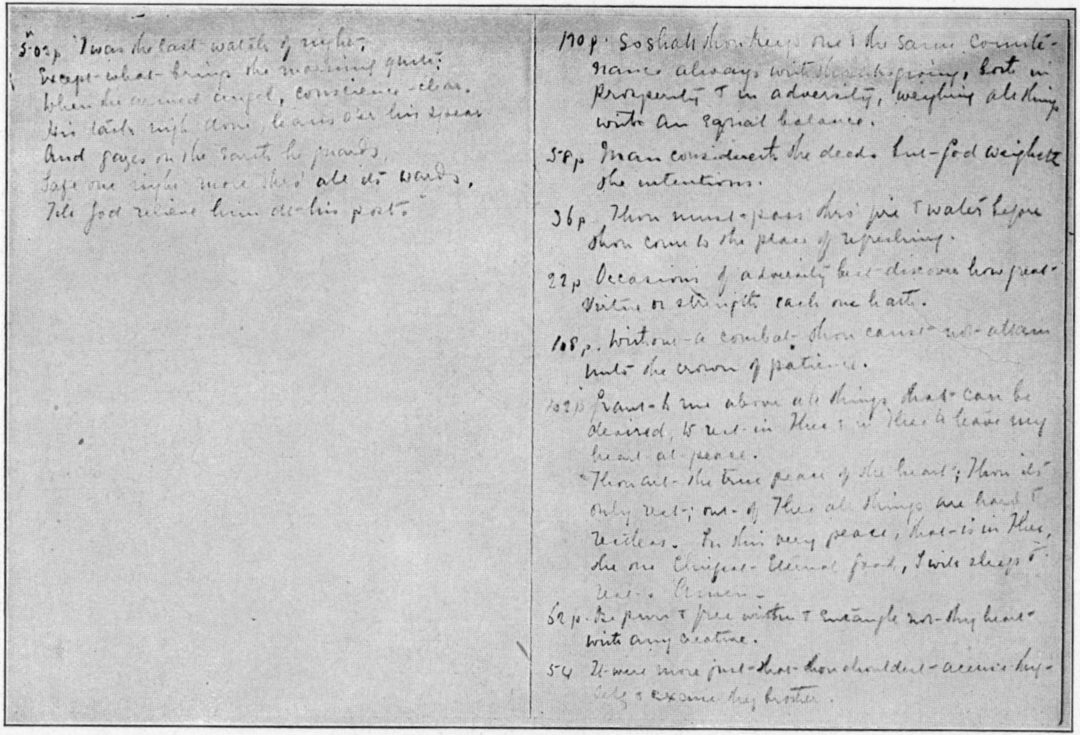

5·0 p 'Twas the last watch of night,
Except what brings the morning quite;
When the armed angel, conscience-clear,
His task nigh done, leans o'er his spear
And gazes on the earth he guards,
Safe one night more thro' all its wards,
Till God relieve him at his post.

170 p. So shalt thou keep one & the same countenance always with thanksgiving, both in prosperity & in adversity, weighing all things with an equal balance.

58 p Man considereth the deeds, but God weigheth the intentions.

36 p Thou must pass thro' fire & water before thou come to the place of refreshing.

22 p Occasions of adversity best discover how great virtue or strength each one hath.

108 p. Without a combat thou canst not attain unto the crown of patience.

132 p Grant to me above all things that can be desired, to rest in Thee & in Thee to leave my heart at peace.
Thou art the true peace of the heart; Thou its only rest; out of Thee all things are hard & restless. In this very peace, that is in Thee, the one Chiefest Eternal Good, I will sleep & rest. Amen.

52 p. Be pure & free within & entangle not thy heart with any creature.

54 It were more just that thou shouldest accuse thyself & excuse thy brother.

Notes in Miss Cavell's prayer book

Notes in Miss Cavell's prayer book

We did not stop until after midnight, when it was only too clear that there was no hope.

It was a bitter business leaving the place feeling that we had failed and that the little woman was to be led out before a firing squad within a few hours. But it was worse to go back to the Legation to the little group of English women who were waiting in my office to learn the result of our visit. They had been there for nearly four hours while Mrs. Whitlock and Miss Larner sat with them and tried to sustain them through the hours of waiting. There were Mrs. Gahan, wife of the English chaplain, Miss B., and several nurses from Miss Cavell's school. One was a little wisp of a thing who had been mothered by Miss Cavell, and was nearly beside herself with grief. There was no way of breaking the news to them gently, for they could read the answer in our faces when we came in. All we could do was to give them each a stiff drink of sherry and send them home. De Leval was white as death, and I took him back to his house. I had a splitting head-ache myself and could not face the idea of going to bed. I went home and read for awhile, but that was no good, so I went out and walked the streets, much to the annoy-ance of German patrols. I rang the bells of several houses in a desperate desire to talk to somebody, but could not find a soul—only sleepy and disgruntled ser-vants. It was a night I should not like to go through again, but it wore through somehow and I braced up with a cold bath and went to the Legation for the day's work.

The day brought forth another loathsome fact in connection with the case. It seems the sentence on Miss Cavell was not pronounced in open court. Her executioners, apparently in the hope of concealing their intentions from us, went into her cell and there, behind locked doors, pronounced sentence upon her. It is all of a piece with the other things they have done.

Last night Mr. Gahan got a pass and was admitted to see Miss Cavell shortly before she was taken out and shot. He said she was calm and prepared and faced the ordeal without a tremor. She was a tiny thing that looked as though she could be blown away with a breath, but she had a great spirit. She told Mr. Gahan that soldiers had come to her and asked to be helped to the frontier; that knowing the risks they ran and the risks she took, she had helped them. She said she had nothing to regret, no complaint to make, and that if she had it all to do over again, she would change nothing. And most pathetic of all was her statement that she thanked God for the six weeks she had passed in prison—the nearest approach to rest she had known for years.

They partook together of the Holy Communion, and she who had so little need of preparation was prepared for death. She was free from resentment and said: "I realise that patriotism is not enough. I must have no hatred or bitterness toward any one."

She was taken out and shot before daybreak.

She was denied the support of her own clergyman at the end, but a German military chaplain stayed with her and gave her burial within the precincts of the prison. He did not conceal his admiration and said: "She was courageous to the end. She professed her Christian faith and said that she was glad to die for her country. She died like a heroine."

DATE DUE

X1598	20 Jan '65		
4475	23 Feb '65		
45/20	1 Sep '65		
DEC 22			
NOV 07 1987			
APR 02 1990			
4/23/90			
FEB 11 1997		•	